REVISE OCR AS/A LEVEL

Biology A

REVISION WORKBOOK

Series Consultant: Harry Smith

Authors: Kayan Parker, Colin Pearson and Rebekka Harding-Smith

Reviewer: David Barrett

Our revision resources are the smart choice for those revising for OCR AS/A Level Biology A. This book will help you to:

- **Organise** your revision with the one-topic-per-page format
- **Prepare** for your AS/A Level exam with a book full of exam-style practice questions
- **Simplify** your revision by writing straight into the book just as you would in an exam
- **Track** your progress with at-a-glance check boxes
- **Improve** your understanding, and exam technique, with guided questions to build confidence, and hints to support key revision points.

Revision is more than just this Workbook!

Make sure that you have practised every topic covered in this book, with the accompanying OCR AS/A Level Biology A Revision Guide. It gives you:

- A 1-to-1 page match with this Workbook
- Explanations of key concepts delivered in short memorable chunks
- Key hints and tips to reinforce your learning
- Worked examples showing you how to lay out your answers
- Exam-style practice questions with answers.

Contents

1-to-1 page match with the OCR Biology Revision Guide ISBN 9781447984368

A small bit of small print

OCR publishes Sample Assessment Material and the Specification on its website. This is the official content and this book should be used in conjunction with it. The questions in Now try this have been written to help you practise every topic in the book. Remember: the real exam questions may not look like this.

Using a light microscope

Practical skills

1 Just by looking at a light microscope, describe how you can work out its magnification.

Remember that a light microscope has two lenses.

....

....

....

.... **(2 marks)**

Guided

2 Which of these statements are true and which are false?

Think back to times when you have used a light microscope. What did you observe and what did it look like?

	True	False
(a) Light microscopes can be used to observe dead specimens.	☐	☒
(b) Light microscopes can be used to observe tissues in colour.	☐	☐
(c) Light microscopes can be used to observe organelles of a cell.	☐	☐
(d) Light microscopes can be used toobserve the movement of cells.	☐	☐

(3 marks)

3 (a) What features of a plant cell would you expect to be able to see using a light microscope?

....

....

....

....

....

.... **(3 marks)**

(b) Explain why, when using a light microscope, you can only see the features you mentioned in part (a).

Discuss magnification and resolution.

....

....

....

.... **(2 marks)**

Using other microscopes

Practical skills

1 Which one of these statements is true for a scanning electron microscope?

☐ **A** Scanning electron microscopes can be used to observe 2D images.

☐ **B** Scanning electron microscopes can be used to observe cell surfaces.

☐ **C** Scanning electron microscopes can be used to observe cells in colour.

☐ **D** Scanning electron microscopes can be used to observe cells in real time. **(1 mark)**

Guided

Maths skills

2 What is the difference between magnification and resolution?

Your definition of magnification needs to refer to both the size of the image and the size of the object.

Magnification is the number of times larger ....

....

Resolution is the ability to distinguish between ....

....

The higher the resolution, the ....

.... **(3 marks)**

3 *Evaluate the strengths and weaknesses of electron and light microscopy.

Aim to give a detailed evaluation using your knowledge of biological procedures. Give evidence to support your points if possible. Make sure your answer is clear and logically structured and that your points are linked

....

....

....

....

....

....

....

....

....

.... **(6 marks)**

Preparing microscope slides

Practical skills

1 The figure shows a transverse section of the stem of a typical pondweed. Part of a graticule is shown below the stem.
The distance between 0 and 10 on the graticule is 1 mm.

(a) Measure the width of the stem between points A and B as it appears on the printed diagram.

.... **(1 mark)**

Maths skills

(b) Work out the width of the stem in real life, using the graticule. Give your answer to the nearest 0.1 mm.

Width of stem = ... **(2 marks)**

2 Which of these statements give uses of staining in light microscopy?

Statement 1: Stains can be used to distinguish between different cell types.

Statement 2: Stains can be used to fix specimens.

Statement 3: Stains can be used to distinguish between different tissue types.

Statement 4: Stains can be used to highlight particular organelles.

☐ **A** 1 and 3 only

☐ **B** 2 and 4 only

☐ **C** 1, 3 and 4 only

☐ **D** 2, 3 and 4 only **(1 mark)**

3 Describe how to prepare a temporary mount of a specimen for light microscopy.

....

....

....

....

.... **(3 marks)**

Had a go ☐ Nearly there ☐ Nailed it! ☐

Calculating magnification

Maths skills

1 Which formula would you use to calculate the actual size of a cell?

☐ **A** magnification divided by image size

☐ **B** image size divided by magnification

☐ **C** magnification multiplied by image size

☐ **D** actual size multiplied by image size

(1 mark)

Maths skills

2 The figure shows a eukaryotic cell.

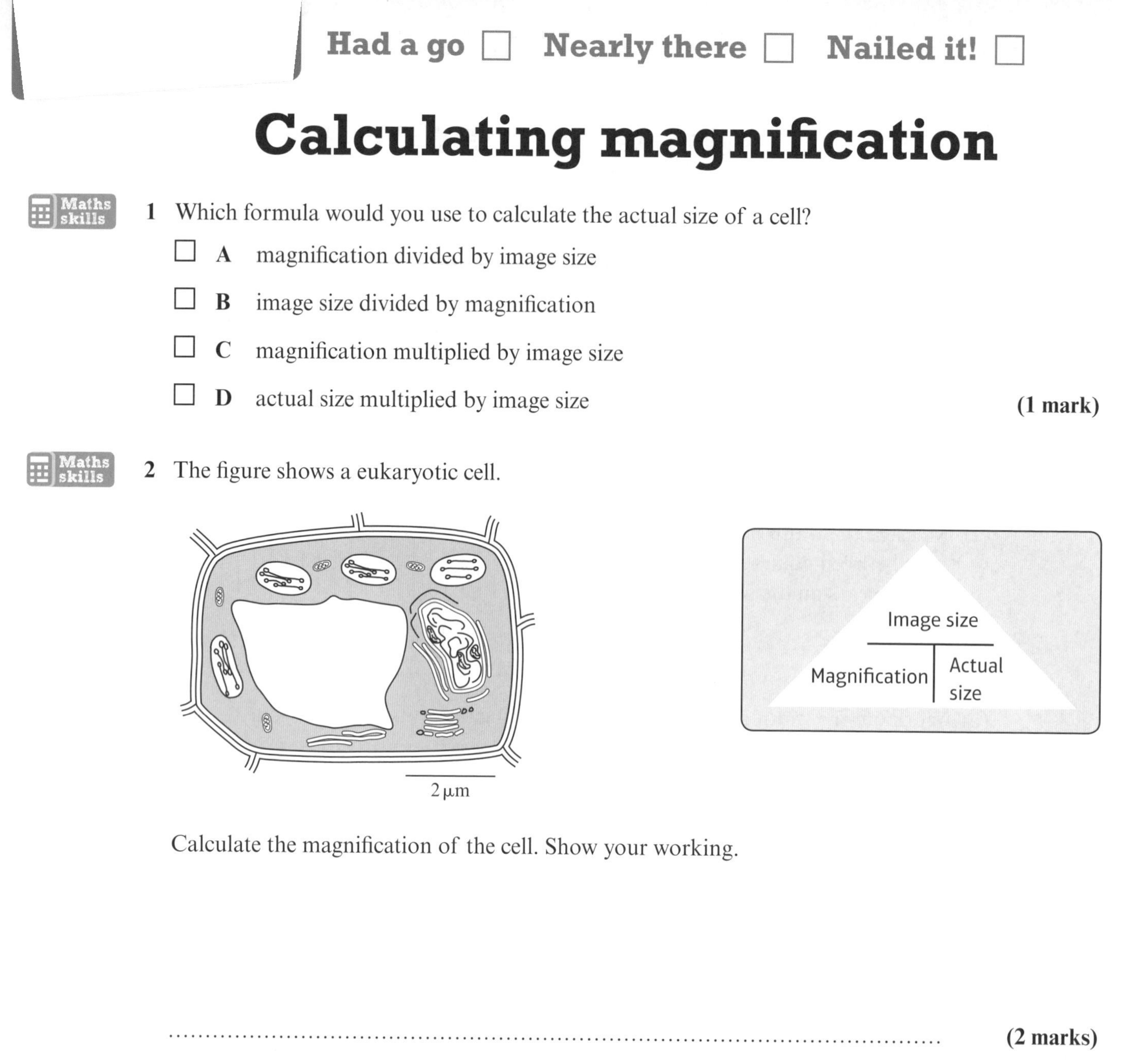

Calculate the magnification of the cell. Show your working.

.... **(2 marks)**

3 The figure is an electron micrograph showing part of a nucleus.

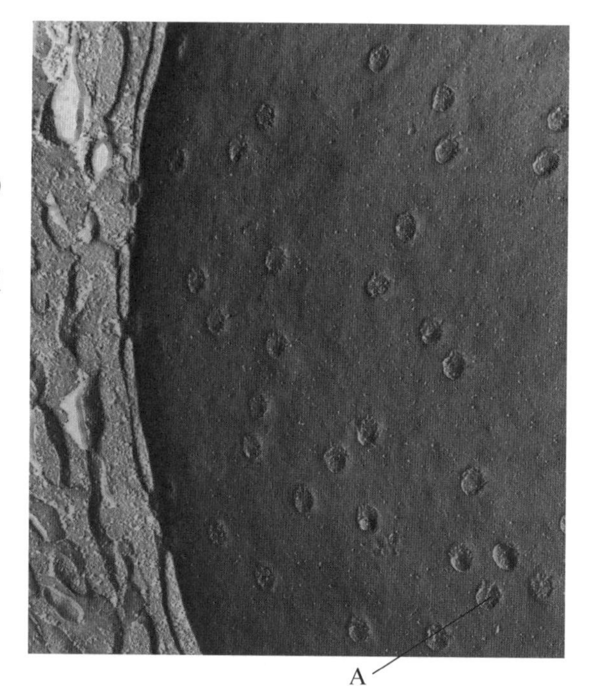

(a) What is A?

.... **(1 mark)**

Maths skills

(b) If the image size of A is 4 mm at a magnification of ×25 000, calculate the actual size of A. Show your working.

.... **(3 marks)**

Eukaryotic and prokaryotic cells

Guided

1 Complete the table by putting a tick or a cross into each column.

	Present in animal cells	Present in plant cells
nucleus	✓	✓
nucleolus		
rough endoplasmic reticulum		
smooth endoplasmic reticulum		
ribosomes		
mitochondria		
chloroplasts		
Golgi apparatus		
centrioles		
cellulose cell wall		
plasma membrane		
large central vacuole		
cytoskeleton		

(6 marks)

2 Mark the following sentences as true or false. Put a tick into the correct box.

	True	False
(a) Prokaryotic cells do not have a nucleus.	☐	☐
(b) Prokaryotic cells have many organelles.	☐	☐
(c) Prokaryotic cells have a cellulose cell wall.	☐	☐
(d) Prokaryotic cells have ribosomes.	☐	☐

(4 marks)

3 Describe the differences in function between the rough endoplasmic reticulum, the smooth endoplasmic reticulum and the Golgi apparatus.

....

....

....

....

....

.... **(3 marks)**

The secretion of proteins

Guided

1 Describe the process by which proteins are made and released.

Proteins are made by the ....

Proteins are made in the ....

The proteins are packaged ....

These travel to the ....

This organelle then modifies the proteins and packages them into ....

....

Finally, they are carried to the ....

.... **(6 marks)**

2 Which one of the following statements is true?

☐ **A** Proteins cross the plasma membrane by facilitated diffusion.

☐ **B** Proteins cross the plasma membrane by diffusion.

☐ **C** Proteins cross the plasma membrane by exocytosis.

☐ **D** Proteins cross the plasma membrane by osmosis. **(1 mark)**

3 (a) Explain why the beta cells of the pancreas have lots of rough endoplasmic reticulum (ER)

Beta cells produce insulin, a protein.

....

....

....

....

....

.... **(3 marks)**

(b) Suggest why the beta cells of the pancreas need a large amount of ATP.

....

....

....

.... **(2 marks)**

The cytoskeleton

1 Which of the following statements are true?

Statement 1: The cytoskeleton is made of microtubules.

Statement 2: The cytoskeleton is involved in the production of proteins.

Statement 3: Flagella and cilia are made of microfilaments.

Statement 4: Microtubule proteins transport vesicles along microtubules.

☐ **A** only 1 and 3

☐ **B** only 1 and 4

☐ **C** only 2 and 3

☐ **D** only 2, 3 and 4 **(1 mark)**

2 What are the roles of these parts of the cytoskeleton?

(a) Flagella: ....

.... **(1 mark)**

(b) Cilia: ....

.... **(1 mark)**

Guided

(c) Microtubules: role is to maintain the shape of the cell. **(1 mark)**

(d) Microtubule motors: ....

.... **(1 mark)**

3 How is the cytoskeleton involved in movement inside and outside of the cell?

....

....

....

....

....

.... **(3 marks)**

The properties of water

Guided

1 Explain the origin of hydrogen bonds between water molecules.

Water is made of one .... atom and two ....

.... atoms. Water is a ....

molecule, which means that the oxygen atom has a slight ....

.... charge and hydrogen has a slight ....

....

.... **(5 marks)**

2 Which one of the following statements is false?

☐ **A** Ice is denser than water.

☐ **B** Water has thermal stability.

☐ **C** Water is a solvent.

☐ **D** Water molecules are involved in many biological reactions. **(1 mark)**

3 Give **three** important functions of water in mammals.

....

....

....

....

....

.... **(3 marks)**

4 Suggest **one** advantage and **one** disadvantage for organisms living on the polar ice.

....

....

....

.... **(2 marks)**

The biochemistry of life

1 Which of these are commonly used inorganic ions in prosthetic groups?

1 magnesium ion (Mg^{2+})

2 iron ion (Fe^{2+})

3 sodium ion (Na^{+})

4 potassium ion (K^{+})

☐ **A** only 1 and 2

☐ **B** only 2 and 4

☐ **C** only 1, 2 and 3

☐ **D** all of them

(1 mark)

2 Which row correctly identifies the polymers made from these monomers?

	alpha-glucose	**beta-glucose**	**amino acids**	**nucleotides**
A	starch	amylose	polypeptide	polynucleotides
B	polypeptide	polynucleotides	amylose	starch
C	amylose	starch	polypeptide	polynucleotides
D	polynucleotides	amylose	cellulose	polypeptide

Rows: ....

(1 mark)

3 Explain the role of water in making and breaking bonds.

Think about how water is used in forming polypeptides such as polysaccharides.

....

....

....

....

....

....

....

....

(4 marks)

Glucose

1 Which of the following statements is true?

Statement 1: Glucose is a disaccharide.

Statement 2: A glycosidic bond forms between monosaccharides.

Statement 3: Glycosidic bonds are broken by a condensation reaction.

☐ **A** 1, 2 and 3

☐ **B** only 1 and 3

☐ **C** only 2 and 3

☐ **D** only 2 **(1 mark)**

Guided 2 Describe how lactose is broken down into its monosaccharides.

Think about the hydrolysis reaction too.

Lactose is made of one glucose monosaccharide and one

.... monosaccharide and so can be described as a

....

....

....

.... **(3 marks)**

3 Some people are lactose intolerant because they cannot digest lactose. Yoghurt made with live bacteria can be eaten safely by some people who are lactose intolerant. Explain why.

Bacteria can digest lactose.

....

....

....

....

....

....

....

.... **(4 marks)**

Starch, glycogen and cellulose

1 Which of the following statements are true?

Statement 1: Starch and glycogen are both made of alpha glucose.

Statement 2: Starch is the energy storage molecule in animals.

Statement 3: Glycogen is made of amylose and amylopectin.

Statement 4: Cellulose is made of beta glucose.

☐ **A** 1 and 2

☐ **B** 1, 2 and 4

☐ **C** 1 and 4

☐ **D** 2, 3 and 4 **(1 mark)**

Guided

2 Describe how the structure of starch and glycogen makes them easy to digest.

Glucose molecules are joined together by ....

Both starch and glycogen are ....

....

.... **(3 marks)**

3 Describe the structural and functional differences between starch and cellulose.

....

....

....

....

....

....

....

.... **(6 marks)**

Triglycerides and phospholipids

Guided

1 Complete the table showing the similarities and differences between triglycerides and phospholipids.

	Triglyceride	Phospholipid
number of ester bonds	3	
number of fatty acid chains		
presence of a glycerol molecule?		✓
presence of a phosphate group?		

(4 marks)

2 Which of the following statements are true?

Statement 1: Ester bonds form between glycerol and a fatty acid chain.

Statement 2: Fatty acid chains are always saturated.

Statement 3: Lipids can act as hormones.

Statement 4: The formation of an ester bond is a hydrolysis reaction.

☐ **A** only 1 and 4

☐ **B** only 1 and 3

☐ **C** only 2, 3 and 4

☐ **D** only 1, 3 and 4

(1 mark)

3 Draw a labelled diagram to show the formation of an ester bond in a triglyceride.

Draw the glycerol molecule and at least one fatty acid. Only the detail of the end of the fatty acid is required.

(3 marks)

Use of lipids in living organisms

1 Which of the following are uses of triglycerides?

Statement 1: insulation in animals

Statement 2: energy storage in animals

Statement 3: steroid hormones

Statement 4: part of the plasma membrane

☐ **A** only 1, 2 and 4

☐ **B** only 2, 3 and 4

☐ **C** only 1 and 2

☐ **D** only 3 and 4

(1 mark)

2 Describe and explain the role of phospholipids, glycolipids and cholesterol in the plasma membrane.

For 'Describe and explain questions', you need to describe what is requested and give reasons or say why you have come to that conclusion.

....

....

....

....

....

....

....

.... **(4 marks)**

3 Explain the uses of triglycerides in aquatic mammals.

....

....

....

....

....

.... **(3 marks)**

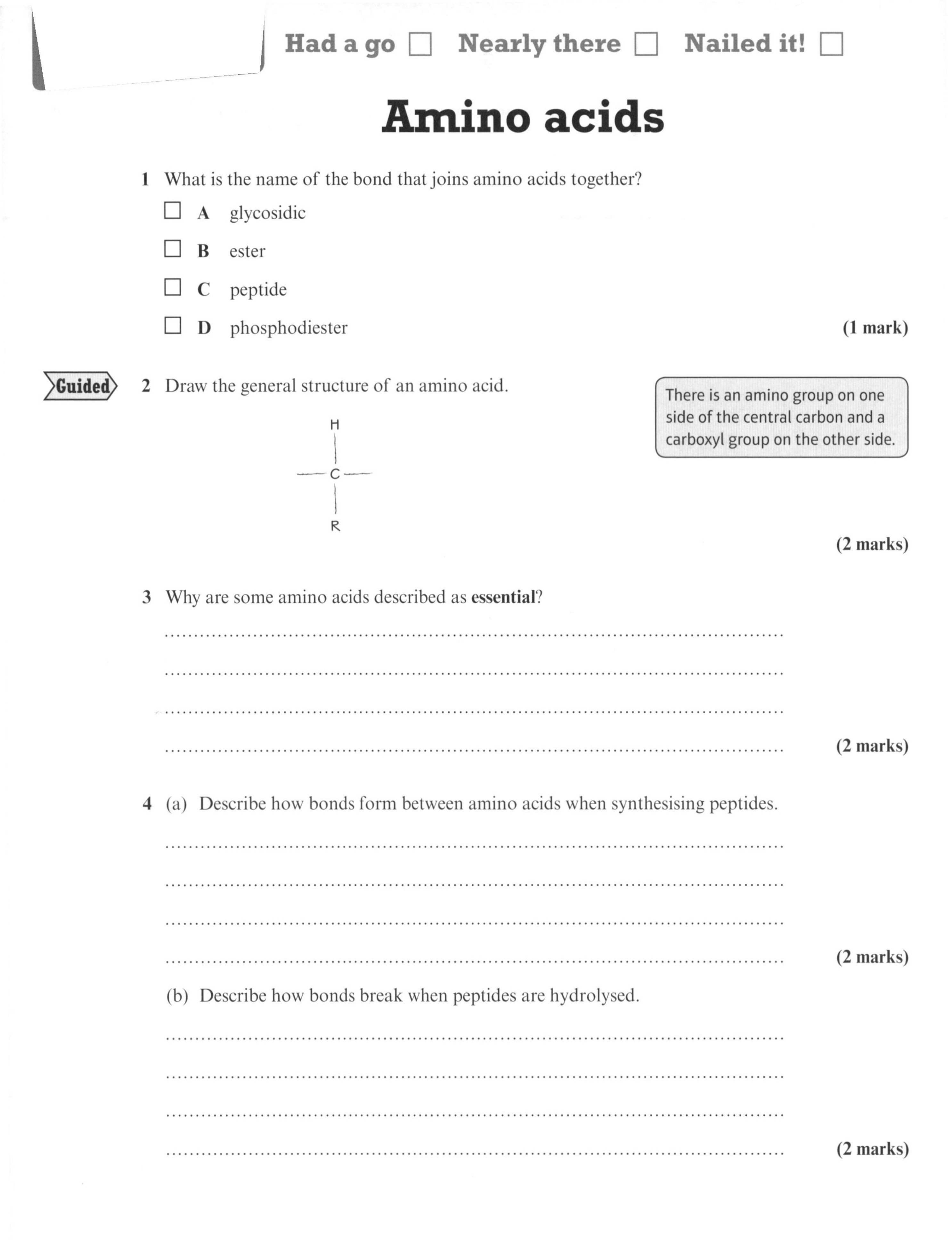

Amino acids

1 What is the name of the bond that joins amino acids together?

☐ **A** glycosidic

☐ **B** ester

☐ **C** peptide

☐ **D** phosphodiester **(1 mark)**

Guided

2 Draw the general structure of an amino acid.

There is an amino group on one side of the central carbon and a carboxyl group on the other side.

(2 marks)

3 Why are some amino acids described as **essential**?

....

....

....

.... **(2 marks)**

4 (a) Describe how bonds form between amino acids when synthesising peptides.

....

....

....

.... **(2 marks)**

(b) Describe how bonds break when peptides are hydrolysed.

....

....

....

.... **(2 marks)**

Protein

1 (a) Compare and contrast the bonding involved in the primary and secondary structure of a protein.

....

....

....

.... **(2 marks)**

(b) Compare and contrast the tertiary and quaternary structure of a protein.

You must discuss both similarities and differences in compare and contrast questions and look at both angles.

....

....

....

.... **(2 marks)**

2 Which of the following proteins has a quaternary structure?

☐ **A** myoglobin

☐ **B** collagen

☐ **C** insulin

☐ **D** amylase **(1 mark)**

3 *Amino acids have different R groups. Describe how these R groups can interact to form the tertiary structure of a protein.

R groups may contain sulfur or carry a positive or negative charge.

....

....

....

....

....

....

....

....

....

....

.... **(6 marks)**

Fibrous proteins

1 *Discuss the functions of fibrous proteins, with examples.

Discuss **three** fibrous proteins, with an example for each.

....

(6 marks)

2 Compare the structure of globular and fibrous proteins.

Discuss both similarities and differences.

....

(3 marks)

3 Outline how fibrous proteins are used in the skeletal system.

Restrict the outline to essential detail only.

....

(2 marks)

Globular proteins

1 What is a **prosthetic group**?

....

.... **(1 mark)**

2 Which of the following is a property of globular proteins?

☐ **A** soluble in water

☐ **B** strong

☐ **C** hard

☐ **D** elastic **(1 mark)**

3 Haemoglobin is a globular protein found in red blood cells.

(a) Name a feature of haemoglobin that makes it a useful component of red blood cells.

....

.... **(1 mark)**

Guided

(b) Describe the structure of haemoglobin.

Haemoglobin is made of four polypeptide chains. ....

....

....

....

....

....

....

....

.... **(5 marks)**

(c) Suggest why a lack of iron in a person's diet could lead to a decrease in the amount of oxygen being carried in the blood.

....

.... **(1 mark)**

Benedict's test

Practical skills

1 Put the stages of the Benedict's test for a non-reducing sugar in the correct order.

Note that this question is about a non-reducing sugar. What difference does the type of sugar make to the test?

Stage 1: Add Benedict's reagent.

Stage 2: Boil for 5 minutes.

Stage 3: Neutralise with sodium hydrogencarbonate ($NaHCO_3$).

Stage 4: Add dilute hydrochloric acid (HCl).

Stage 5: Heat at 80 °C for 3 minutes.

☐ **A** 1 5 4 3 2

☐ **B** 4 2 3 1 5

☐ **C** 3 4 1 5 2

☐ **D** 5 4 3 1 2

(1 mark)

Guided

2 Describe how Benedict's reagent identifies a reducing sugar.

Benedict's reagent contains ....

Reducing sugars, such as glucose ....

....

....

.... **(3 marks)**

3 Suggest which test should be used to test for the presence of sugars in orange juice. Explain your answer.

....

....

....

....

....

....

....

.... **(4 marks)**

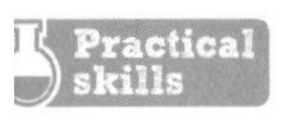

Tests for protein, starch and lipids

1 Fill in the table below detailing the positive and negative results for the following food tests.

Test	Positive	Negative
biuret's test	....	blue
iodine test	....	....
emulsion test	white emulsion	....

(4 marks)

2 Which polymer is the iodine test testing for?

☐ **A** polynucleotide

☐ **B** polypeptide

☐ **C** polypropene

☐ **D** polysaccharide

(1 mark)

3 (a) A sample of milk undergoes the biuret's test, the iodine test and the emulsion test. Explain the results you would expect.

....

....

....

....

....

....

(3 marks)

(b) Explain the results you would expect to see if the iodine test and the emulsion test were carried out on a sample of potato.

....

....

....

....

(2 marks)

Practical techniques – colorimetry

1 What does colorimetry measure?

☐ **A** concentration

☐ **B** density

☐ **C** absorbance

☐ **D** volume **(1 mark)**

Guided

2 Put the stages of carrying out colorimetry in the correct order.

A Use a sample of distilled water to calibrate the colorimeter.

B The concentration of the unknown sample is worked out using the calibration curve.

C The absorbance of each unknown sample is measured.

D The filter in the colorimeter is changed to the colour that would be best absorbed by the sample.

E The absorbances are used to draw a calibration curve.

F The absorbance of each known sample is measured.

G Samples of known concentrations are added to the colorimeter.

..A.... ....E.... .... **(4 marks)**

3 A sample of river water was taken at several places along a river, close to a polluting factory. Describe how colorimetry could be used to test the samples of river water for iron deposits.

Iron turns water a red/brown colour.

....

....

....

....

....

....

....

.... **(4 marks)**

Practical skills

Practical techniques – chromatography

1 Which of the following statements are true for paper chromatography?

Statement 1: Chromatography is a quantitative method.

Statement 2: Chromatography separates a mixture depending on the differential affinities of the components.

Statement 3: Chromatography uses a solvent to separate a mixture.

Statement 4: The R_f value of each component can be calculated using the position of the solvent front and the distance travelled by each component.

☐ **A** 1, 2 and 3

☐ **B** 2, 3 and 4

☐ **C** 1, 3 and 4

☐ **D** 1, 2 and 4

(1 mark)

Guided

Maths skills

2 (a) The table gives some results from two chromatography experiments, using the same solvent and stationary phase. Calculate the missing values.

Sample	Solute front (cm)	Solvent front (cm)	R_f value
A	1.5	15.0	0.1
B	6.0	15.0	....
C	4.0	12.0	....
D	4.8	12.0	....

How to work out Rf value:

$R_f \text{ value} = \frac{\text{solute front}}{\text{solvent front}}$

How to work out percentage increase:

Percentage increase

$= \frac{(\text{end value} - \text{start value})}{\text{start value}} \times 100\%$

(3 marks)

(b) Use these results to suggest which samples could be the same compound.

....

....

(1 mark)

3 Describe how to carry out paper chromatography on a sample of ink.

....

....

....

....

....

....

....

....

....

....

(5 marks)

Exam skills

1 The figure below shows a plant cell.

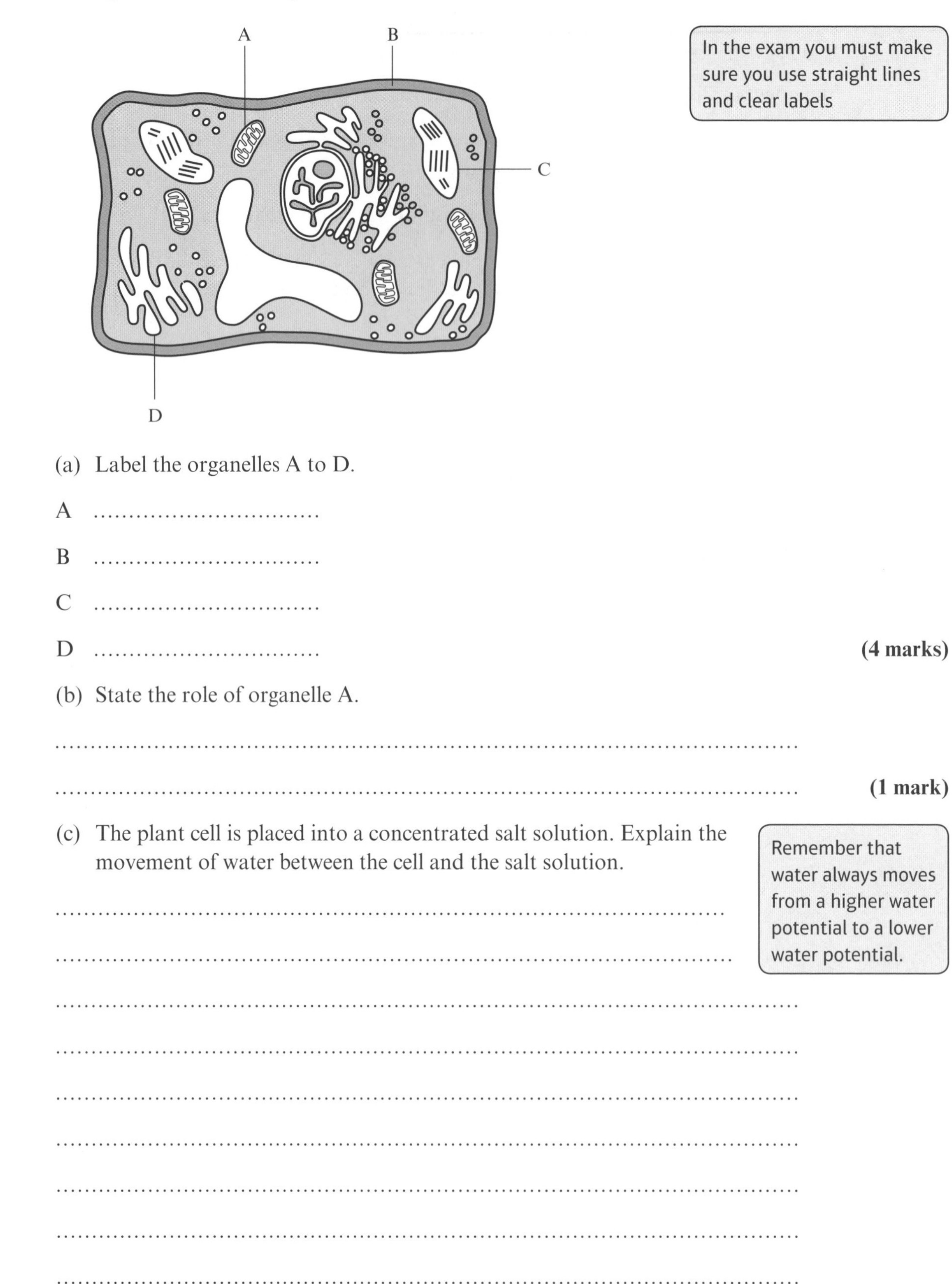

In the exam you must make sure you use straight lines and clear labels

(a) Label the organelles A to D.

A ....

B ....

C ....

D .... **(4 marks)**

(b) State the role of organelle A.

....

.... **(1 mark)**

(c) The plant cell is placed into a concentrated salt solution. Explain the movement of water between the cell and the salt solution.

Remember that water always moves from a higher water potential to a lower water potential.

....

....

....

....

....

....

....

....

....

.... **(4 marks)**

1 The figure shows part of a DNA molecule.

(a) Identify X, Y and Z.

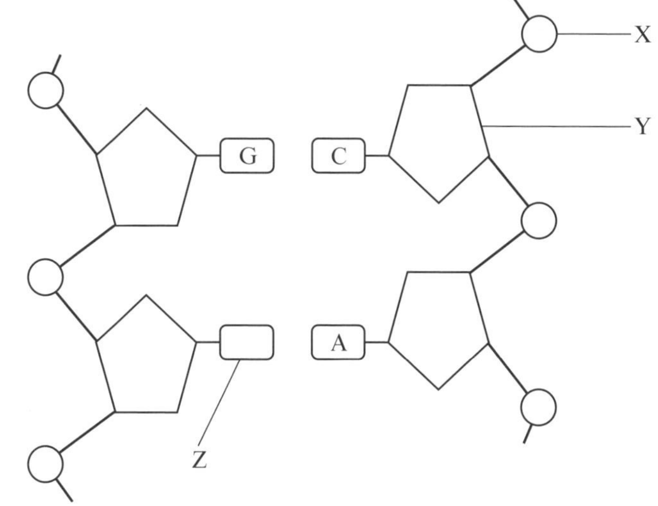

X ....

Y ....

Z ....

(3 marks)

(b) What type of bond is found between X and Y in the backbone of the DNA molecule?

☐ **A** phosphodiester bond

☐ **B** ionic bond

☐ **C** peptide bond

☐ **D** hydrogen bond **(1 mark)**

2 RNA nucleotides form molecules in the nucleus and the cytoplasm of the cell.

(a) What type of RNA molecule is found within the nucleus?

....

.... **(1 mark)**

(b) Explain how bonds are formed between RNA nucleotides.

....

....

....

....

....

.... **(3 marks)**

(c) What type of reaction breaks the bonds between RNA nucleotides?

....

.... **(1 mark)**

ADP and ATP

1 What are the components of an ATP molecule?

☐ **A** adenine, deoxyribose sugar and two phosphate groups

☐ **B** adenine, deoxyribose sugar and three phosphate groups

☐ **C** adenine, ribose sugar and two phosphate groups

☐ **D** adenine, ribose sugar and three phosphate groups **(1 mark)**

Guided

2 (a) Describe the similarities and differences between a molecule of ATP and an RNA nucleotide.

ATP and RNA nucleotides both contain ....

....

and ....

....

ATP and RNA nucleotides contain different numbers of phosphate

groups: ....

.... **(3 marks)**

(b) Describe how ATP is used in the body.

....

....

....

.... **(2 marks)**

(c) Explain how energy is released from ATP.

....

....

....

....

....

.... **(3 marks)**

The structure of DNA

1 Which of the following statements is correct?

☐ **A** DNA nucleotides are joined together by hydrogen bonds.

☐ **B** DNA is made of two parallel strands of DNA that twist to form a double helix.

☐ **C** A purine base always pairs with a pyrimidine base.

☐ **D** DNA contains a hexose sugar called ribose. **(1 mark)**

2 The sequence of nitrogenous bases on one DNA strand is TACCTTCATTTAGGG.

(a) What is the complementary DNA sequence?

.... **(1 mark)**

(b) Describe how the nitrogenous bases on each DNA strand pair up.

....

....

....

....

....

.... **(3 marks)**

Practical skills

3 DNA was removed from some cheek cells. Describe the process of precipitating DNA from the cells.

DNA needs to be removed from the nucleus, which is surrounded by a nuclear envelope.

....

....

....

....

....

.... **(3 marks)**

Semi-conservative DNA replication

1 Which enzyme catalyses the breaking of hydrogen bonds between the nitrogenous bases on the DNA strands?

☐ **A** DNA ligase

☐ **B** DNA helicase

☐ **C** DNA polymerase

☐ **D** restriction endonuclease **(1 mark)**

2 (a) What is meant by **semi-conservative** DNA replication?

....

....

....

.... **(2 marks)**

(b) Describe the process of DNA replication.

You could include some of the enzymes named in question 1 in your answer.

....

....

....

....

....

.... **(3 marks)**

(c) Explain the consequences if DNA is not copied exactly during each replication.

....

....

....

....

....

.... **(3 marks)**

The genetic code

1 Describe the genetic code.

Use the words 'degenerate', 'universal' and 'non-overlapping' in your answer.

....

....

....

....

....

.... **(3 marks)**

2 Use the table to work out the amino acid sequence of the following genetic code: ATGGGTTGTTATCACTGA.

.... **(1 mark)**

Each set of three bases is one triplet. The three letter names in bold are the abbreviated names of amino acids, e.g. **Phe** = phenylalanine.

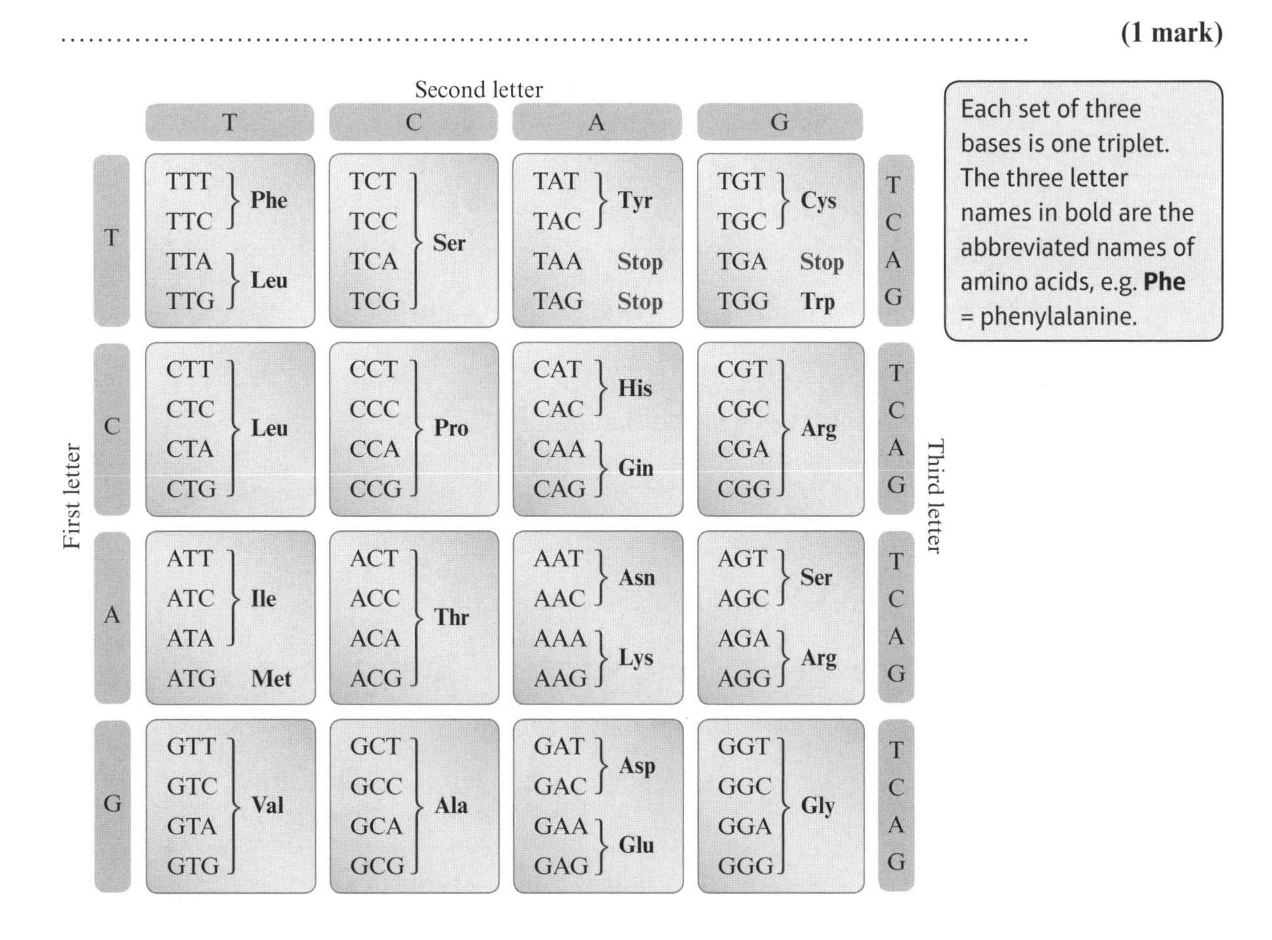

First letter	Second letter: T	C	A	G	Third letter
T	TTT **Phe**	TCT **Ser**	TAT **Tyr**	TGT **Cys**	T
T	TTC **Phe**	TCC **Ser**	TAC **Tyr**	TGC **Cys**	C
T	TTA **Leu**	TCA **Ser**	TAA **Stop**	TGA **Stop**	A
T	TTG **Leu**	TCG **Ser**	TAG **Stop**	TGG **Trp**	G
C	CTT **Leu**	CCT **Pro**	CAT **His**	CGT **Arg**	T
C	CTC **Leu**	CCC **Pro**	CAC **His**	CGC **Arg**	C
C	CTA **Leu**	CCA **Pro**	CAA **Gin**	CGA **Arg**	A
C	CTG **Leu**	CCG **Pro**	CAG **Gin**	CGG **Arg**	G
A	ATT **Ile**	ACT **Thr**	AAT **Asn**	AGT **Ser**	T
A	ATC **Ile**	ACC **Thr**	AAC **Asn**	AGC **Ser**	C
A	ATA **Ile**	ACA **Thr**	AAA **Lys**	AGA **Arg**	A
A	ATG **Met**	ACG **Thr**	AAG **Lys**	AGG **Arg**	G
G	GTT **Val**	GCT **Ala**	GAT **Asp**	GGT **Gly**	T
G	GTC **Val**	GCC **Ala**	GAC **Asp**	GGC **Gly**	C
G	GTA **Val**	GCA **Ala**	GAA **Glu**	GGA **Gly**	A
G	GTG **Val**	GCG **Ala**	GAG **Glu**	GGG **Gly**	G

3 Changes in the base sequence of genes are called mutations. What effect could a deletion mutation have on the function of a protein?

....

....

....

....

....

.... **(3 marks)**

Transcription and translation of genes

1 A DNA template has the nucleotide sequence ATGCGCATATCCCAA. Which row below shows the DNA sequence that is complementary to the DNA template and shows the mRNA sequence of the DNA template?

		Complementary DNA sequence	mRNA sequence
☐	**A**	TACGCGTATAGGGTT	UACGCGUAUAGGGUU
☐	**B**	CATGTGCACAGGGCC	CAUGUGCACAGGGCC
☐	**C**	UACGCGUAUAGGGUU	ATGCGCATATCCCAA
☐	**D**	CAUGUGCACAGGGCC	GTACACGTGTCCCGG

(1 mark)

2 (a) Explain how messenger RNA is transcribed from DNA.

Include the use of enzymes in transcription.

....

....

....

....

....

.... **(3 marks)**

(b) What is the role of tRNA in translation?

....

....

....

....

....

....

....

....

....

.... **(5 marks)**

The role of enzymes

1 Which one of the following statements is true?

☐ **A** Enzymes are non-biological catalysts that speed up the rate of reactions.

☐ **B** Enzymes carry out metabolic reactions within cells and are used up by these reactions.

☐ **C** Extracellular enzymes are made inside cells and are secreted by exocytosis.

☐ **D** Enzymes work by increasing the activation energy in a reaction. **(1 mark)**

2 (a) (i) What is meant by the term **intracellular** when referring to enzymes?

....

.... **(1 mark)**

(ii) Give **one** example of an intracellular enzyme.

.... **(1 mark)**

(b) Trypsin in an extracellular enzyme. Explain how trypsin is secreted into the small intestine.

Recall that the Golgi apparatus is involved in the secretion of proteins.

....

....

....

....

....

.... **(3 marks)**

3 Amylase is an enzyme found in saliva, and has an optimum pH of 7. Assess what would happen to this enzyme when it reached the stomach (pH 1).

Assess questions require you to identify important factors, make a judgement of importance and reach a conclusion.

....

....

....

.... **(3 marks)**

4 Explain why living organisms use enzymes.

....

....

....

.... **(3 marks)**

The mechanism of enzyme action

1 Explain the **lock and key** mechanism of enzyme action.

....

....

....

....

....

.... **(3 marks)**

2 Which of the following statements are correct?

Statement 1: The substrate fits into the active site of the enzyme in a complementary way.

Statement 2: Each enzyme is specific for its substrate.

Statement 3: Enzymes are fibrous proteins.

Statement 4: When the enzyme is bound to the product, it is called the enzyme–product complex.

☐ **A** 1 and 2 only

☐ **B** 1, 2 and 3 only

☐ **C** 1, 2 and 4 only

☐ **D** all of the above **(1 mark)**

3 Sucrase is an enzyme that breaks down sucrose. Describe what happens to sucrose after it binds to the active site of sucrase.

Use the terms 'enzyme–substrate complex' and 'enzyme–product complex'.

....

....

....

....

....

.... **(3 marks)**

Factors that affect enzyme action

1 Which of the following graphs shows the change in the rate of reaction of a fixed volume of enzyme as the substrate concentration increases? Tick the correct box.

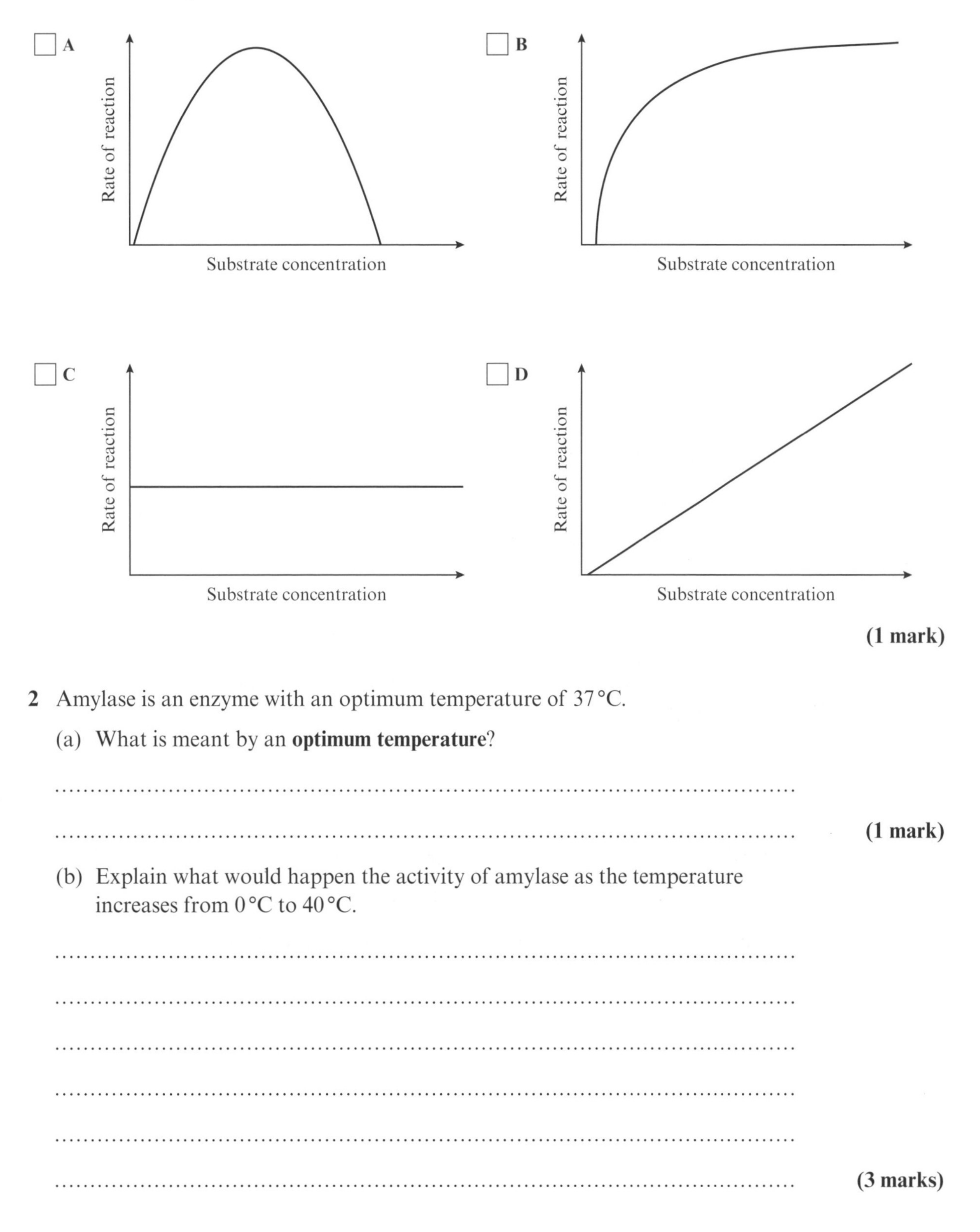

(1 mark)

2 Amylase is an enzyme with an optimum temperature of 37 °C.

(a) What is meant by an **optimum temperature**?

....

.... **(1 mark)**

(b) Explain what would happen the activity of amylase as the temperature increases from 0 °C to 40 °C.

....

....

....

....

....

.... **(3 marks)**

Factors that affect enzyme action – practical investigations

Practical skills

1 An investigation was carried out into the activity of catalase on hydrogen peroxide. The results are shown in the graph.

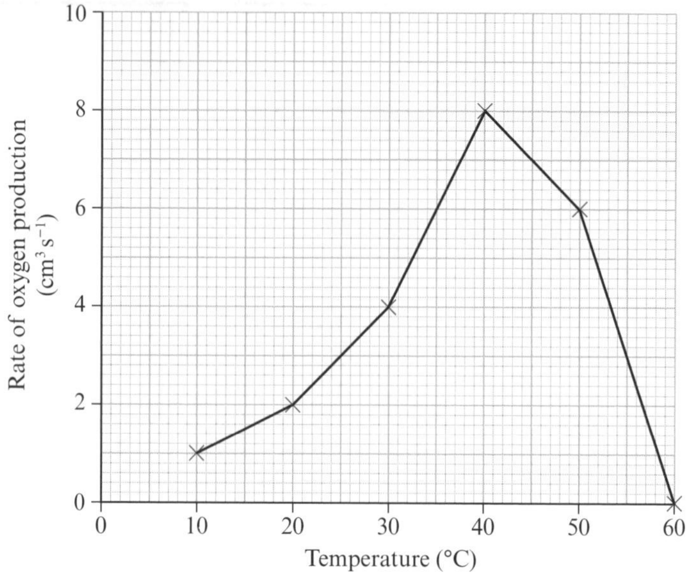

(a) What is the optimum temperature?

.... **(1 mark)**

Maths skills

(b) What is the rate of oxygen production at the optimum temperature?

....

.... **(1 mark)**

Maths skills

(c) What is the percentage increase in the rate of oxygen production between 20 °C and 40 °C?

....

....

....

.... **(2 marks)**

(d) Explain the change in the rate of oxygen production after 40 °C.

....

....

....

....

....

.... **(3 marks)**

(e) What is the temperature coefficient of catalase between 20 °C and 30 °C?

☐ **A** 2 ☐ **B** 4 ☐ **C** 8 ☐ **D** 10 **(1 mark)**

Cofactors, coenzymes and prosthetic groups

1 Which of the following statements is true?

☐ **A** A prosthetic group is a type of cofactor.

☐ **B** A coenzyme is not used up by the reaction.

☐ **C** An enzyme without its cofactor is called a holoenzyme.

☐ **D** Co-substrates are a type of prosthetic group. **(1 mark)**

Guided

2 (a) What is a cofactor?

Cofactors are non-protein inorganic substances. ....

....

.... **(1 mark)**

(b) Explain how a co-substrate can aid the formation of an enzyme–substrate complex.

....

....

....

.... **(2 marks)**

3 (a) Give **two** examples of prosthetic groups, and describe their role within the protein.

....

....

....

.... **(2 marks)**

(b) Explain why minerals and vitamins are important for enzyme and protein action in the human body.

....

....

....

.... **(2 marks)**

Think about what coenzymes and prosthetic groups are made from.

Inhibitors

1 The graph shows the rate of reaction of an enzyme with and without an inhibitor.

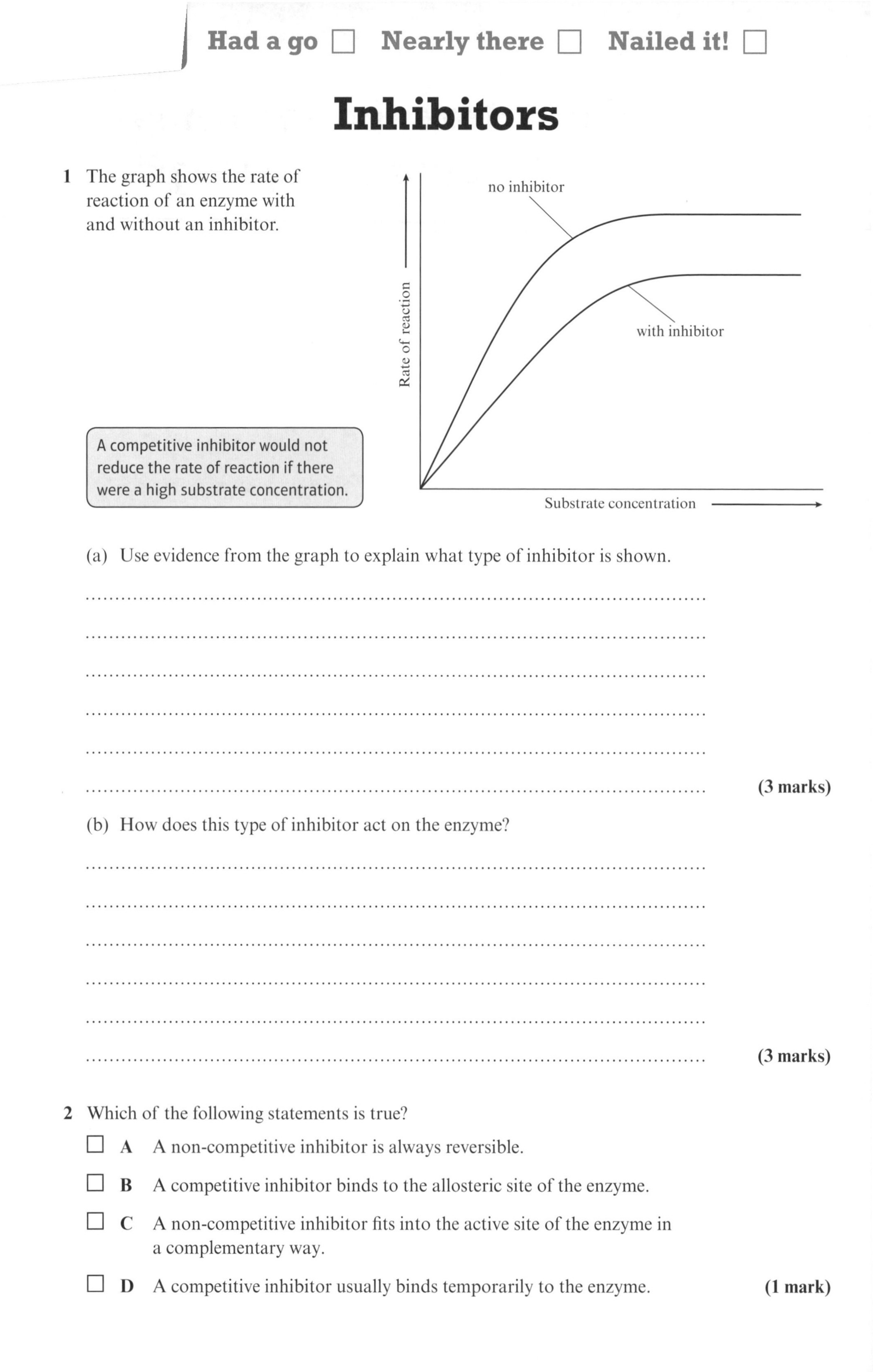

A competitive inhibitor would not reduce the rate of reaction if there were a high substrate concentration.

(a) Use evidence from the graph to explain what type of inhibitor is shown.

....

....

....

....

....

.... **(3 marks)**

(b) How does this type of inhibitor act on the enzyme?

....

....

....

....

....

.... **(3 marks)**

2 Which of the following statements is true?

☐ **A** A non-competitive inhibitor is always reversible.

☐ **B** A competitive inhibitor binds to the allosteric site of the enzyme.

☐ **C** A non-competitive inhibitor fits into the active site of the enzyme in a complementary way.

☐ **D** A competitive inhibitor usually binds temporarily to the enzyme. **(1 mark)**

Exam skills

1 (a) Draw an alpha glucose molecule.

The formula for glucose is $C_6H_{12}O_6$. Make sure you have all of these atoms in your diagram.

(2 marks)

(b) Alpha glucose molecules join together in long polysaccharide chains to form glycogen. What is the name of the bond between two alpha glucose molecules?

.... **(1 mark)**

(c) Compare and contrast the structure of glycogen and starch.

....

....

....

....

....

....

....

.... **(4 marks)**

(d) A test is carried out to see if a sample contains glucose. Describe the test.

....

....

....

....

....

.... **(3 marks)**

(e) The test result is green. Explain what this means.

....

.... **(1 mark)**

The fluid mosaic model

1 Which of the following are components of the plasma membrane?

Component 1: glycoprotein

Component 2: cholesterol

Component 3: phospholipid

Component 4: glycolipid

☐ **A** 1, 2 and 3 only

☐ **B** 1, 2 and 4 only

☐ **C** 2, 3 and 4 only

☐ **D** all of the above

(1 mark)

2 Integral proteins are found in the plasma membrane.

(a) What does **integral** mean?

....

.... **(1 mark)**

Guided

(b) Name the **two** different types of integral protein and explain their roles.

Channel protein: ....

....

....

Carrier protein: ....

....

.... **(2 marks)**

3 Describe how the plasma membrane allows hormones to affect the cell.

Include the role of receptor proteins.

....

....

....

....

....

.... **(3 marks)**

Factors that affect membrane structure

1 Describe the role of cholesterol in the plasma membrane.

....

....

....

.... **(2 marks)**

2 What is the effect of a solvent on a plasma membrane?

☐ **A** denatures plasma membranes

☐ **B** dissolves plasma membranes

☐ **C** immobilises the plasma membrane

☐ **D** increases the kinetic energy of the plasma membrane **(1 mark)**

3 Yeast cells were placed into water at different temperatures. The water contained methylene blue. After a fixed period of time, the percentage of cells containing methylene blue was measured. The results are shown in the table.

Temperature (°C)	Number of cells containing methylene blue (%)
0	0
15	12
30	14
45	24
60	54

(a) Use the information in the table to explain how temperature affects the uptake of methylene blue.

....

....

....

.... **(2 marks)**

Guided

(b) Suggest why the data in the table show that the methylene blue is not actively transported into the yeast cells.

As the temperature increases above 40 °C, carrier proteins would start to denature. ....

....

....

.... **(2 marks)**

Movement across the membrane

1 Which process allows lipids to cross the plasma membrane into the cell?

☐ **A** active transport

☐ **B** facilitated diffusion

☐ **C** osmosis

☐ **D** diffusion **(1 mark)**

2 (a) Describe how glucose is transported into the cell.

Mention **two** mechanisms for glucose to cross the membrane.

....

....

....

....

....

....

....

.... **(4 marks)**

(b) Explain why glucose cannot enter the cell by simple diffusion.

....

....

....

.... **(2 marks)**

3 Plasma cells produce large volumes of antibodies for use in the immune response.

Antibodies are proteins, and are too large to pass through a channel or carrier protein.

(a) Which process transports the antibodies out of the cell?

....

.... **(1 mark)**

(b) How is this process different to phagocytosis?

....

....

....

.... **(2 marks)**

Osmosis

1 (a) Describe the process of **osmosis**.

....

....

....

....

....

.... **(3 marks)**

(b) What is the effect of osmosis on plant cells placed into a solution of high salt concentration?

....

....

....

.... **(2 marks)**

Guided

2 (a) What is meant by the term **water potential**?

Water potential is a measure of ....

....

It is a type of ....

.... **(2 marks)**

(b) Suggest why it is important for plant cells that the water potential inside the cells is more negative than the water potential outside the cells.

....

....

....

.... **(2 marks)**

3 An animal cell with a water potential of −200 kPa is placed into pure water. What will happen to the cell?

Think about the type of cell.

☐ **A** Water will move into the cell, causing the cell to burst.

☐ **B** Water will move out of the cell, causing the cell to shrivel.

☐ **C** Water will move into the cell, causing the cell to become turgid.

☐ **D** Water will move out of the cell, causing the cell to become plasmolysed. **(1 mark)**

Had a go ☐ Nearly there ☐ Nailed it! ☐

Factors affecting diffusion

Maths skills

1 A cube is 3 cm in width, depth and length. What is its surface area to volume ratio?

☐ **A** 6:1

☐ **B** 3:1

☐ **C** 2:1

☐ **D** 1.5:1 **(1 mark)**

Practical skills

2 An experiment was carried out to determine the sugar concentration of cell cytoplasm. Potato chips were weighed, placed into solutions of different sugar concentrations for a time, and then weighed again. Some potato chips gained mass, whilst others lost mass.

(a) Why did the potato chips gain or lose mass?

Use your knowledge of osmosis to answer part (a).

....

....

....

.... **(2 marks)**

(b) Explain how the sugar concentration of the cell cytoplasm of the potato chips was identified.

....

....

....

....

....

.... **(3 marks)**

Guided

(c) (i) Name **three** variables that must be kept constant in this investigation.

The mass of the potato chips,

.... **(3 marks)**

(ii) Suggest **two** potential sources of error in this investigation.

....

....

....

.... **(2 marks)**

The cell cycle

1 The cell cycle is made up of several stages. Which stage is responsible for the replication of chromosomes?

☐ **A** S phase

☐ **B** G2 phase

☐ **C** mitosis

☐ **D** G1 phase

(1 mark)

2 (a) What term is used to describe the number of times a cell can go through the cell cycle?

....

(1 mark)

(b) Name **two** types of cell that can go through the cell cycle an infinite number of times.

....

....

(2 marks)

(c) Suggest what could happen to a cell if it were prevented from going through the cell cycle.

Cells are prevented from going through the cell cycle at the end of their life or if the cell is damaged.

....

....

....

....

(2 marks)

Guided

3 (a) Describe how the cell cycle is regulated.

The cell cycle is regulated by a regulatory protein, called p53, which acts at the G1/S checkpoint and the G2/M checkpoint.

....

....

....

....

(2 marks)

(b) Explain why it is important to regulate the cell cycle.

....

....

....

....

(2 marks)

Mitosis

1 Which of the following images shows a cell in metaphase?

A **B** **C** **D**

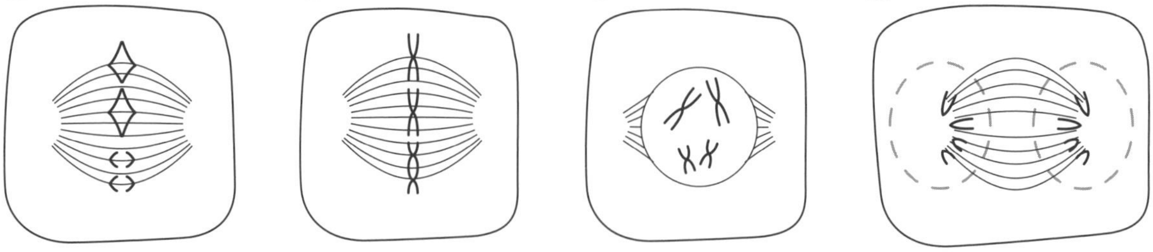

.... **(1 mark)**

2 Mitosis is the cell division of diploid cells.

(a) What is meant by the term **diploid**?

....

.... **(1 mark)**

(b) Describe what happens to the following parts of the cell during prophase.

Nuclear envelope: ....

....

Centrioles: ....

....

Chromosomes: ....

.... **(3 marks)**

3 (a) Describe the use of mitosis in plants.

Include asexual reproduction in your answer.

....

....

....

....

....

.... **(3 marks)**

Practical skills

(b) Suggest a part of a plant that would be good to use in an experiment designed to observe mitosis.

.... **(1 mark)**

Meiosis

1 Which of the following statements are true?

Statement 1: Meiosis produces four genetically identical daughter cells.

Statement 2: Meiosis goes through two rounds of cell division.

Statement 3: Meiosis produces gametes.

Statement 4: Meiosis increases genetic variation by the independent assortment of chromosomes.

☐ **A** 1, 2 and 3 only

☐ **B** 1, 2 and 4 only

☐ **C** 2, 3 and 4 only

☐ **D** All of the above **(1 mark)**

2 What is the order of the stages of the first division of meiosis?

Do you have an mnemonic to help you remember PMAT?

....

....

....

.... **(4 marks)**

3 A diploid cell contains three pairs of homologous chromosomes.

(a) In which stage of meiosis do the homologous chromosomes line up on the equator of the cell?

.... **(1 mark)**

(b) Draw the position of the chromosomes inside the cell during anaphase II.

(2 marks)

(c) Describe how the homologous chromosomes exchange genetic material.

Include the terms 'cross over' and 'chiasma' in your answer.

....

....

....

....

....

.... **(3 marks)**

Specialised cells

1 What are the properties of a neutrophil?

☐ **A** large surface area and packed with haemoglobin

☐ **B** lobed nuclei and packed with lysosomes

☐ **C** large surface area and no chloroplasts

☐ **D** thick cellulose cell wall and work in pairs **(1 mark)**

2 Compare and contrast the features of guard cells and mesophyll palisade cells in the leaves of plants.

This means you need to identify similarities and differences.

....

....

....

....

....

....

....

.... **(4 marks)**

3 (a) Describe how sperm cells are specialised for their function.

....

....

....

....

....

.... **(3 marks)**

(b) Why is it important that sperm cells contain only one set of chromosomes?

....

.... **(1 mark)**

Specialised tissues

1 What is a tissue?

☐ **A** a group of identical cells working together to perform a particular function

☐ **B** a layer of identical cells working together to perform a particular function

☐ **C** a group of similar cells working together to perform a particular function

☐ **D** a layer of several different cells working together to perform a particular function **(1 mark)**

2 (a) Describe how specialised animal tissues are used in the movement of muscles and mucus.

....

....

....

.... **(2 marks)**

Guided (b) Compare and contrast the features of ciliated and squamous epithelium.

Squamous and ciliated epithelium both line inside surfaces.

....

....

....

....

....

....

....

....

.... **(4 marks)**

3 Describe how water and sucrose are moved around a plant.

Include xylem and phloem in your answer.

....

....

....

....

....

....

....

.... **(4 marks)**

Stem cells

1 What are adult stem cells called?

☐ **A** multipotent stem cells

☐ **B** totipotent stem cells

☐ **C** pluripotent stem cells

☐ **D** omnipotent stem cells **(1 mark)**

2 (a) What is a blastocyst?

....

....

....

.... **(2 marks)**

(b) Describe how blastocysts are used in stem cell research.

....

....

....

.... **(2 marks)**

3 (a) Where in a plant are stem cells found?

Stem cells are found where growth occurs in plants.

....

....

....

....

....

.... **(3 marks)**

(b) Explain how new plant tissue is formed from stem cells.

....

....

....

.... **(2 marks)**

Uses of stem cells

1 Which of the following are uses of stem cells?

i growing new organs or parts of organs for transplant

ii growing new skin tissue to replace damaged skin

iii growing new neurones to treat neurological disorders

iv studying the development of the embryo

☐ **A** i, ii and iii only ☐ **C** ii, iii and iv only

☐ **B** i, iii and iv only ☐ **D** all of the above **(1 mark)**

2 Stem cells can be found in embryos, umbilical cord blood and adult tissue.

(a) What are the advantages of using embryonic stem cells instead of stem cells from umbilical cord blood and adult tissue?

....

....

....

....

....

.... **(3 marks)**

(b) Suggest an ethical issue arising from using embryonic stem cells.

....

.... **(1 mark)**

(c) What are **induced pluripotent stem cells**?

....

.... **(1 mark)**

3 (a) Suggest how stem cells could be used to grow new organs.

Here you need to draw on your biological knowledge to answer these questions.

....

....

....

.... **(2 marks)**

(b) Suggest a reason for using adult stem cells to grow organs.

Think about organ transplants.

....

.... **(1 mark)**

Exam skills

1 The diagram below shows the main vein in a leaf.

(a) Label the top and bottom parts of the vein.

(i) ....

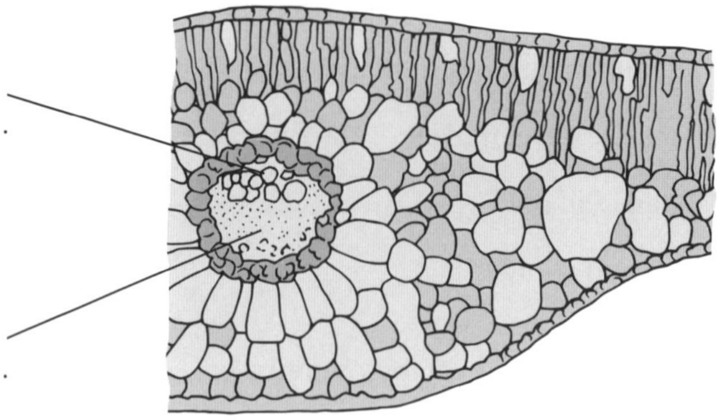

(ii) ....

(2 marks)

(b) What **two** types of cell is phloem tissue made from?

....

....

(2 marks)

(c) Compare and contrast the structure of xylem and phloem tissue.

Think about the shape of the end plates and what makes up the cell walls.

....

....

....

....

....

....

....

....

(4 marks)

(d) A ring was cut into the bark of a tree during the summer and, after a time, the area above the cut bulged out. Explain how this shows the movement of sugars through a plant.

....

....

....

....

(2 marks)

Gas exchange surfaces

Maths skills

1 Imagine a cube with volume 1 cm^3.

(a) Calculate its total surface area.

....

.... **(1 mark)**

Now imagine the same volume 'squashed' so it forms a cuboid with dimensions 4 cm × 4 cm × 0.0625 cm.

Calculate the area (length x width) of all six faces of the cube/cuboid and add them together.

(b) Calculate the new total surface area.

....

.... **(1 mark)**

(c) Which shape would make the better exchange surface? Explain your answer.

....

....

....

....

....

.... **(3 marks)**

2 (a) What size organisms need specialised gas exchange surfaces?

....

.... **(1 mark)**

(b) Why does a specialised gas exchange surface need a rich blood supply?

Think about the time taken for gases to diffuse into the centre of the organism.

....

....

....

.... **(2 marks)**

(c) Describe the specialised gas exchange surface in plants.

....

....

....

.... **(2 marks)**

The lungs

1 This is a cross-section through a bronchus.

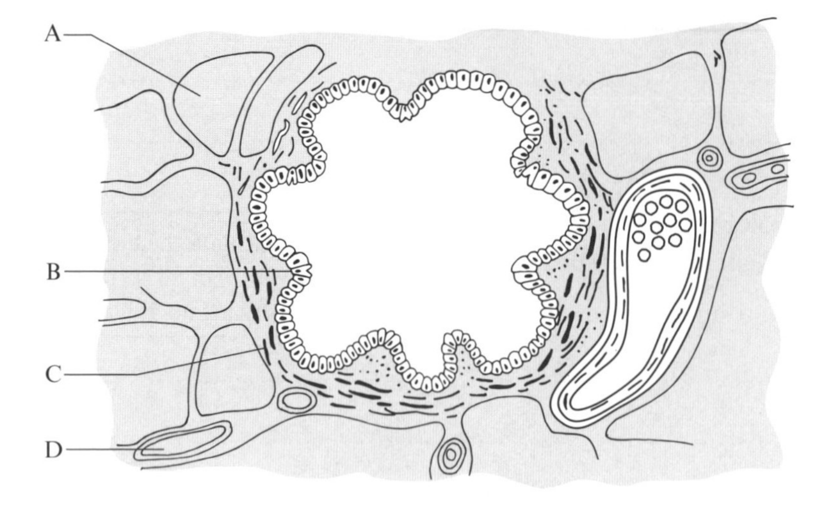

(a) Which tissue contains goblet cells?

.... **(1 mark)**

(b) What is the function of the smooth muscle in the wall of the bronchus?

....

....

....

.... **(2 marks)**

(c) How are blood capillaries involved in gas exchange?

....

....

....

....

....

.... **(3 marks)**

2 Inflammation caused by cigarette smoke causes enzymes to break down the elastic fibres of the alveoli. This can lead to a condition called emphysema.

> Use your knowledge of the surface area of the alveoli in your answer.

Suggest how the structure of the alveoli of a person with emphysema is different to that of a person without emphysema.

....

....

....

.... **(2 marks)**

The mechanism of ventilation

1 Which of the following statements are true?

Statement 1: During expiration, the volume of the thorax increases.

Statement 2: During inspiration, the pressure in the thorax decreases.

Statement 3: During expiration, the external intercostal muscles relax.

Statement 4: During inspiration, the diaphragm becomes dome shaped.

☐ **A** 1 and 2 only

☐ **B** 2 and 3 only

☐ **C** 1 and 4 only

☐ **D** 3 and 4 only **(1 mark)**

2 (a) Describe the process of inspiration during ventilation.

....

....

....

....

....

....

....

.... **(4 marks)**

(b) Explain why it is important to change the pressure inside the thorax during expiration.

Air moves from high pressure to low pressure.

....

....

....

.... **(2 marks)**

3 Explain why not all of the air can be removed from the lungs.

....

....

....

.... **(2 marks)**

Using a spirometer

Maths skills

1 A person's oxygen uptake is 0.5 dm^3 of oxygen in 15 seconds. What is their rate of oxygen uptake?

☐ **A** $0.125\,dm^3\,min^{-1}$

☐ **B** $7.5\,dm^3\,min^{-1}$

☐ **C** $0.5\,dm^3\,min^{-1}$

☐ **D** $2.0\,dm^3\,min^{-1}$ **(1 mark)**

Practical skills

2 The graph below shows a spirometer trace.

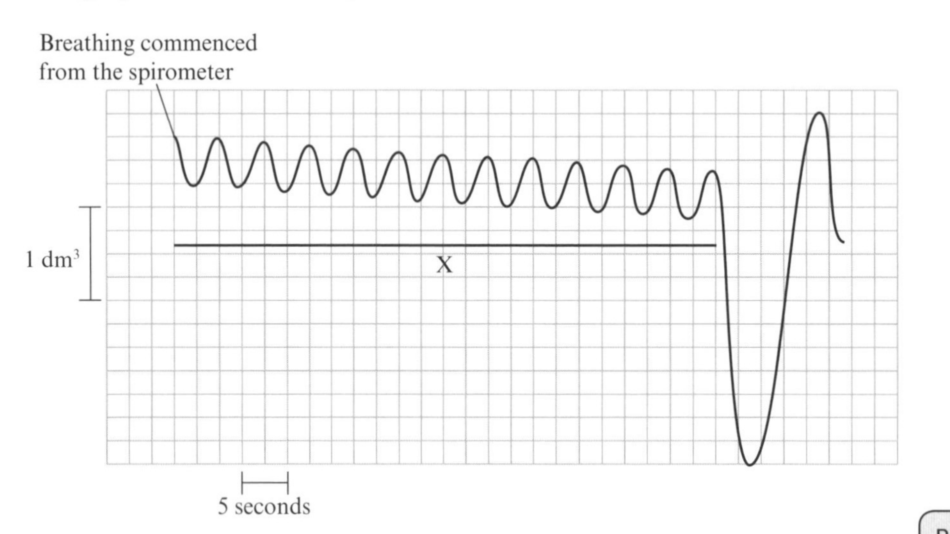

(a) What is the breathing rate during time X?

Breathing rate is measured in breaths per minute. You could check your answer by measuring your breathing rate now. Are the rates similar?

....

....

....

.... **(2 marks)**

(b) Measure the person's vital capacity.

Read the vital capacity from the graph.

....

.... **(1 mark)**

(c) Explain why the peaks and troughs during X go diagonally down the trace.

This 'explain' question requires you to set out reasons for the line of the graph using your knowledge of spirometers.

....

....

....

....

....

.... **(3 marks)**

Ventilation in bony fish and insects

1 In insects, blood flows out of the blood vessels and into the body cavity. What word describes this type of circulatory system?

☐ **A** closed ☐ **C** open

☐ **B** single ☐ **D** double **(1 mark)**

2 Describe the movement of blood and water in the gills of a bony fish.

....

....

....

.... **(2 marks)**

3 (a) Label the parts of this insect gas exchange system.

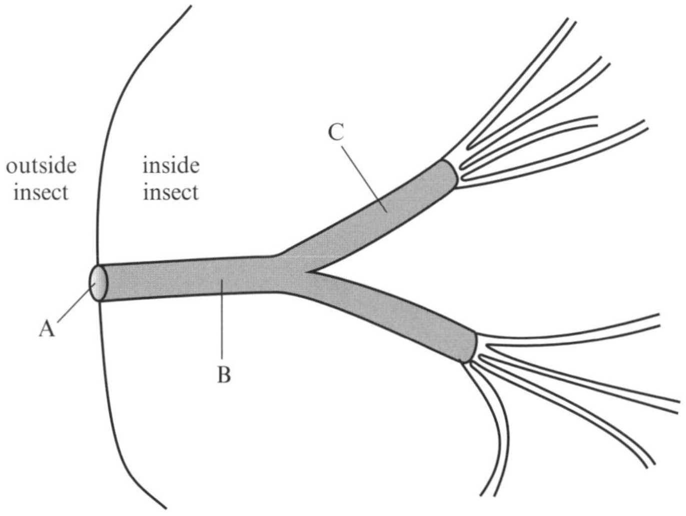

Use the words 'spiracle', 'trachea' and 'tracheole'. Remember that you must use straight lines and clear labels in the exam.

A ....

B ....

C .... **(3 marks)**

(b) Describe how insects take oxygen into their bodies.

For 'describe questions', you do not need to make a judgement or explain how or why, you only need to describe what is requested.

....

....

....

....

....

.... **(3 marks)**

Had a go ☐ Nearly there ☐ Nailed it! ☐

Circulatory systems

1 What is the oxygen-carrying fluid inside the circulatory system of insects called?

☐ **A** haemolymph

☐ **B** haemoglobin

☐ **C** interstitial fluid

☐ **D** tracheal fluid

(1 mark)

2 (a) Describe how single-celled organisms take in oxygen.

....

....

....

.... **(2 marks)**

(b) Explain why it is not necessary for single-celled organisms to have a specialised gas exchange system.

Include the phrase 'surface area to volume ratio' in your answer.

....

....

....

.... **(2 marks)**

3 (a) Compare and contrast the heart of a fish and a human heart.

....

....

....

....

....

.... **(3 marks)**

(b) What is the advantage of having a closed circulatory system?

....

.... **(1 mark)**

Blood vessels

1 Choose the correct order of the tissues in an artery, starting from the inside.

☐ **A** elastic fibres, endothelium, smooth muscle and collagen fibres

☐ **B** smooth muscle, collagen fibres, endothelium and elastic fibres

☐ **C** collagen fibres, smooth muscle, elastic fibres and endothelium

☐ **D** endothelium, elastic fibres, smooth muscle and collagen fibres **(1 mark)**

2 (a) What type of cells are the endothelium tissues made from?

....

.... **(1 mark)**

(b) What is the role of the elastic fibres in the arteries?

....

....

....

.... **(2 marks)**

3 (a) Describe the structural features of arteries and give the function of each.

As this is a three-mark question, name at least **three** different features.

....

....

....

....

....

....

.... **(3 marks)**

(b) How is the blood able to flow back to the heart at low pressure?

....

.... **(2 marks)**

The formation of tissue fluid

Maths skills

1 In a capillary, the hydrostatic pressure is 3.9 kPa and the overall pressure is 2.7 kPa.

Oncotic pressure = overall pressure – hydrostatic pressure

(a) What is meant by the term **oncotic pressure**?

....

.... **(1 mark)**

(b) What is the oncotic pressure in this capillary?

....

.... **(1 mark)**

(c) Explain in which direction the fluid will move.

....

....

....

.... **(2 marks)**

Guided

(d) What is the advantage of the fluid moving in this way?

The fluid, which contains high concentrations of oxygen and glucose,

has moved closer ....

....

There is less distance for ....

.... **(2 marks)**

2 (a) What is the role of the neutrophils in the blood and body tissues?

....

....

....

.... **(2 marks)**

(b) Explain why erythrocytes are only found in the blood.

....

.... **(1 mark)**

The mammalian heart

Guided

1 Label this diagram of the heart.

A .... vein

B Right ....

C ....

D ....

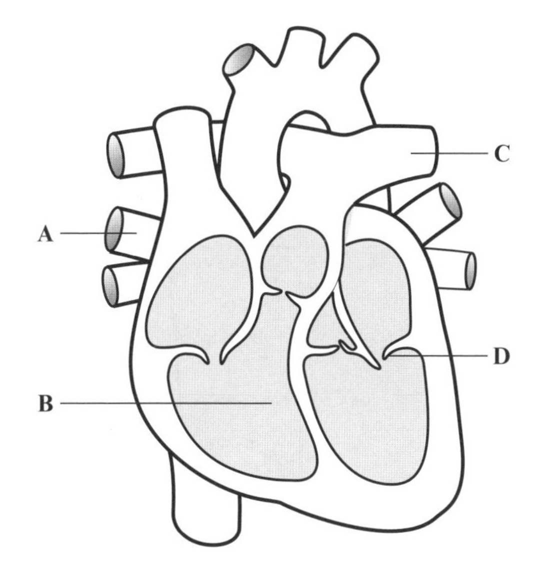

(4 marks)

2 (a) Describe the movement of the blood through the left side of the heart.

....

....

....

....

....

.... **(3 marks)**

(b) Contrast the blood in the left side of the heart with the blood in the right side of the heart.

Think where the blood has come from and where it is going next.

....

....

....

.... **(2 marks)**

3 Explain the difference in thickness of the wall of the right ventricle and the wall of the left ventricle.

....

....

....

....

.... **(3 marks)**

The cardiac cycle

1 At which point on the graph do the semilunar valves close?

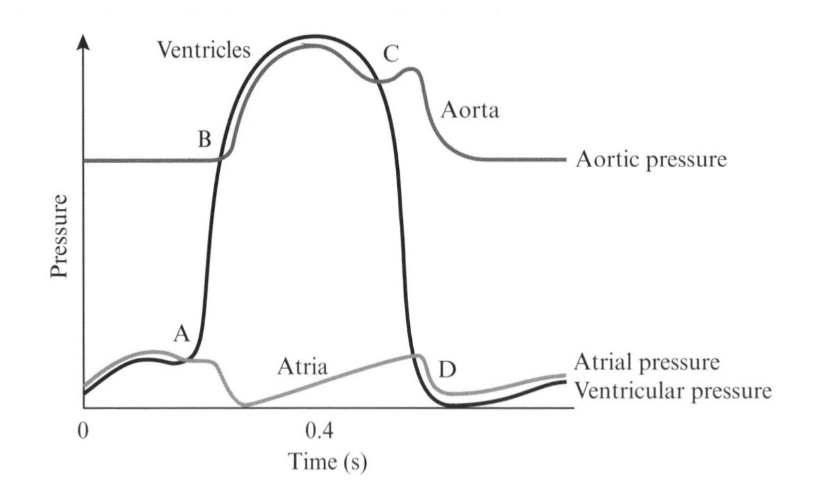

Your answer .... **(1 mark)**

2 (a) Describe how the blood moves from the right atrium to the right ventricle.

> Discuss the change in pressure in the heart chambers.

....

....

....

....

....

.... **(3 marks)**

(b) Explain why the atrioventricular valves close after the blood flows into the ventricles.

....

....

....

.... **(2 marks)**

3 A blood pressure reading is given as 120/80. What does this mean?

....

....

....

.... **(2 marks)**

Control of the heart

1 Which position on the heart indicates the location of the sinoatrial node?

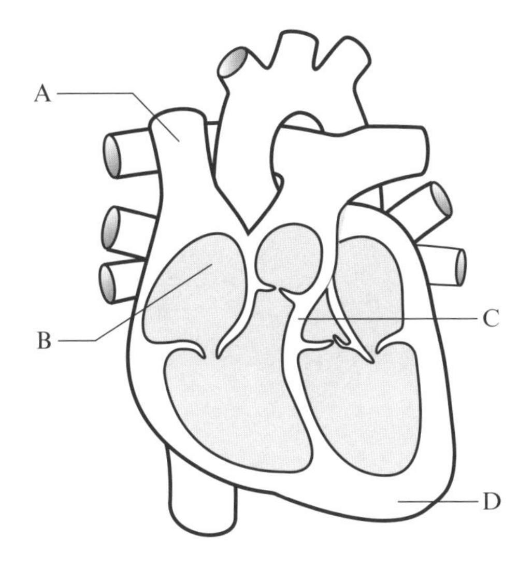

Your answer .... **(1 mark)**

2 Describe how the heartbeat is coordinated.

Include the terms 'sinoatrial node' and 'atrioventricular node' in your answer.

....

....

....

....

....

.... **(3 marks)**

3 The diagram below shows an electrocardiogram (ECG) of the heart.

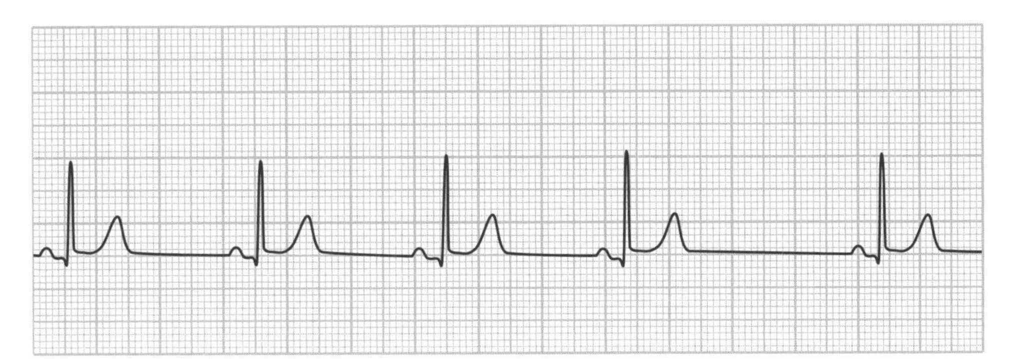

Name the tallest peak. What does it represent?

....

....

....

.... **(2 marks)**

Had a go ☐ Nearly there ☐ Nailed it! ☐

Haemoglobin

Guided

1 Describe how carbon dioxide is carried inside a red blood cell.

About 5% is dissolved ....

....

About 10% combines directly with ....

....

About 85% is transported as ....

....

....

.... **(4 marks)**

2 The diagram below shows the oxygen dissociation curve of adult haemoglobin.

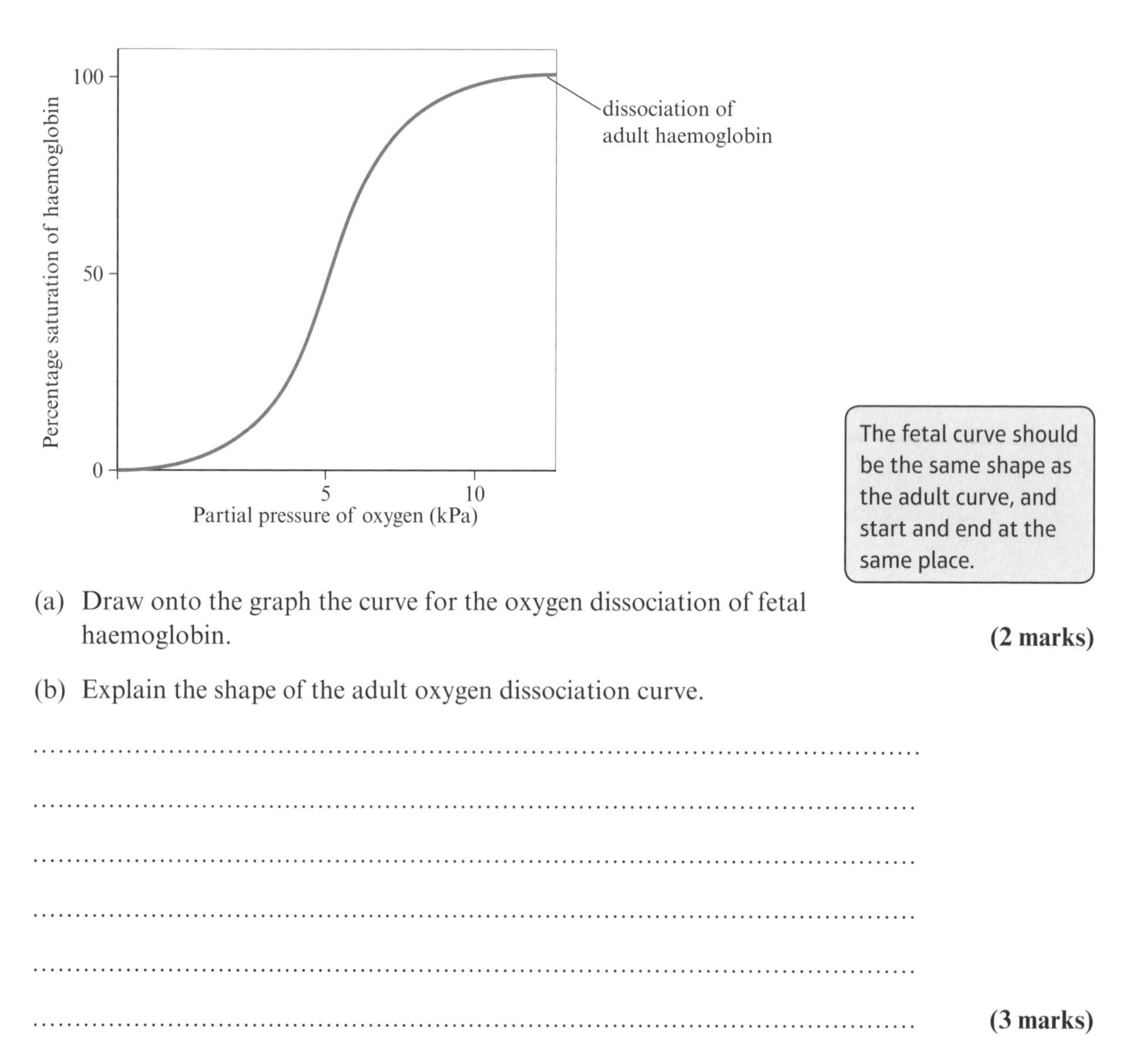

(a) Draw onto the graph the curve for the oxygen dissociation of fetal haemoglobin. **(2 marks)**

(b) Explain the shape of the adult oxygen dissociation curve.

....

....

....

....

....

.... **(3 marks)**

The plant vascular system

1 Which of the following statements is true?

☐ **A** Phloem contains a waterproof layer of lignin in the cell wall.

☐ **B** Xylem is made of many connected cells with sieve plates between them.

☐ **C** Water can move up and down the plant in the xylem, and sideways through pits in the cell wall.

☐ **D** Phloem and companion cells are linked by pores called plasmodesmata. **(1 mark)**

Guided

2 (a) What is xylem tissue made from?

Xylem is made from ....

....

Xylem contains .... which gives strength to

....

.... **(2 marks)**

(b) Explain the advantages of xylem being hollow and narrow.

....

....

....

....

....

.... **(3 marks)**

3 (a) What is the main sugar being transported in the phloem sap?

.... **(1 mark)**

(b) Describe how the cells of the phloem are different to the cells of the xylem.

....

....

....

.... **(2 marks)**

(c) Explain why the activity of the phloem is dependent on ATP.

....

....

....

.... **(2 marks)**

Leaves, stems and roots

1 Name tissues A to D

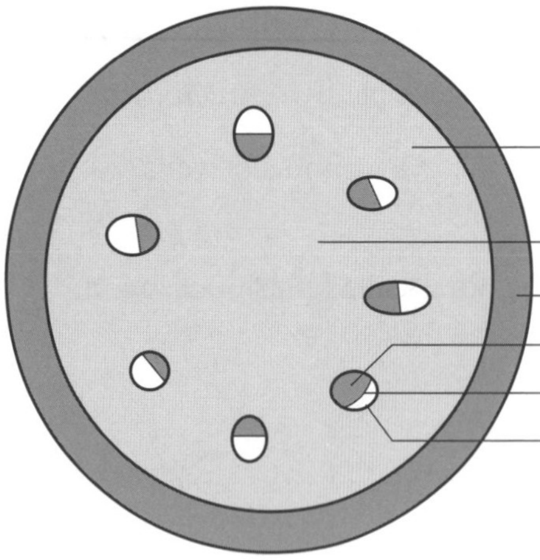

A .... B ....

C .... D .... **(4 marks)**

2 (a) Describe how the xylem and phloem are arranged in a leaf.

....

....

....

.... **(2 marks)**

(b) By what physical process does the water in the xylem move to the palisade cells?

....

.... **(1 mark)**

(c) Describe the role of the companion cells and the phloem in transporting excess sugars from the leaf during the summer.

Use the terms 'source' and 'sink' in your answer.

....

....

....

....

....

....

....

....

....

.... **(5 marks)**

Transpiration

1 Which of the following factors will increase the rate of transpiration?

Factor 1: increased number of stomata open

Factor 2: decreased temperature

Factor 3: decreased humidity

Factor 4: increased wind speed

☐ **A** 1, 2 and 3 only ☐ **C** 2, 3 and 4 only

☐ **B** 1, 3 and 4 only ☐ **D** all of the above. **(1 mark)**

2 (a) Describe the process of transpiration.

Include in your answer the terms 'osmosis', 'evaporates' and 'diffuses'.

....

....

....

....

....

.... **(3 marks)**

(b) Explain why the rate of transpiration is lower during the night than in the day.

....

....

....

.... **(2 marks)**

Practical skills

3 Describe briefly how a bubble potometer can be used to estimate the rate of transpiration.

....

....

....

....

....

.... **(3 marks)**

The transport of water

1 The apoplast pathway is the path of water through:

☐ **A** the cytoplasm of the cells.

☐ **B** the vacuoles of the cells.

☐ **C** the spaces in the cell walls and between the cells.

☐ **D** the Casparian strip in the endodermis. **(1 mark)**

2 Describe the movement of water up the stem of a plant.

Use the terms 'cohesion' and 'adhesion' in your answer.

....

....

....

....

....

....

....

.... **(4 marks)**

3 Explain how hydrophytic plants are adapted to live in water.

....

....

....

....

....

....

....

....

....

.... **(5 marks)**

Translocation

1 Which of the following statements are true?

Statement 1: Sap contains mainly glucose, along with other assimilates.

Statement 2: Components of the phloem sap are actively loaded into the phloem by companion cells.

Statement 3: Excess sugars are stored as starch in the buds of the plant.

Statement 4: Water moves into the phloem by osmosis.

☐ **A** 1 and 2 only

☐ **B** 3 and 4 only

☐ **C** 1 and 3 only

☐ **D** 2 and 4 only **(1 mark)**

Guided

2 (a) Explain how the phloem sap moves down the phloem.

Water moves from the xylem into the top of the phloem by osmosis

....

....

This creates a high pressure at the top of the phloem ....

....

The water potential at the top of the phloem is low because ....

.... **(3 marks)**

(b) What is meant by the term **assimilates**?

....

.... **(1 mark)**

Practical skills

3 Describe evidence for the mass flow theory.

....

....

....

....

....

....

....

.... **(4 marks)**

Had a go ☐ Nearly there ☐ Nailed it! ☐

Exam skills

1 (a) Describe the mechanism of inhalation.

> Describe what happens to the diaphragm and intercostal muscles.

....

....

....

....

....

....

....

.... **(4 marks)**

Maths skills

(b) A person's breathing was measured using a spirometer.

(i) Use the graph below to work out the breathing rate. Show your working.

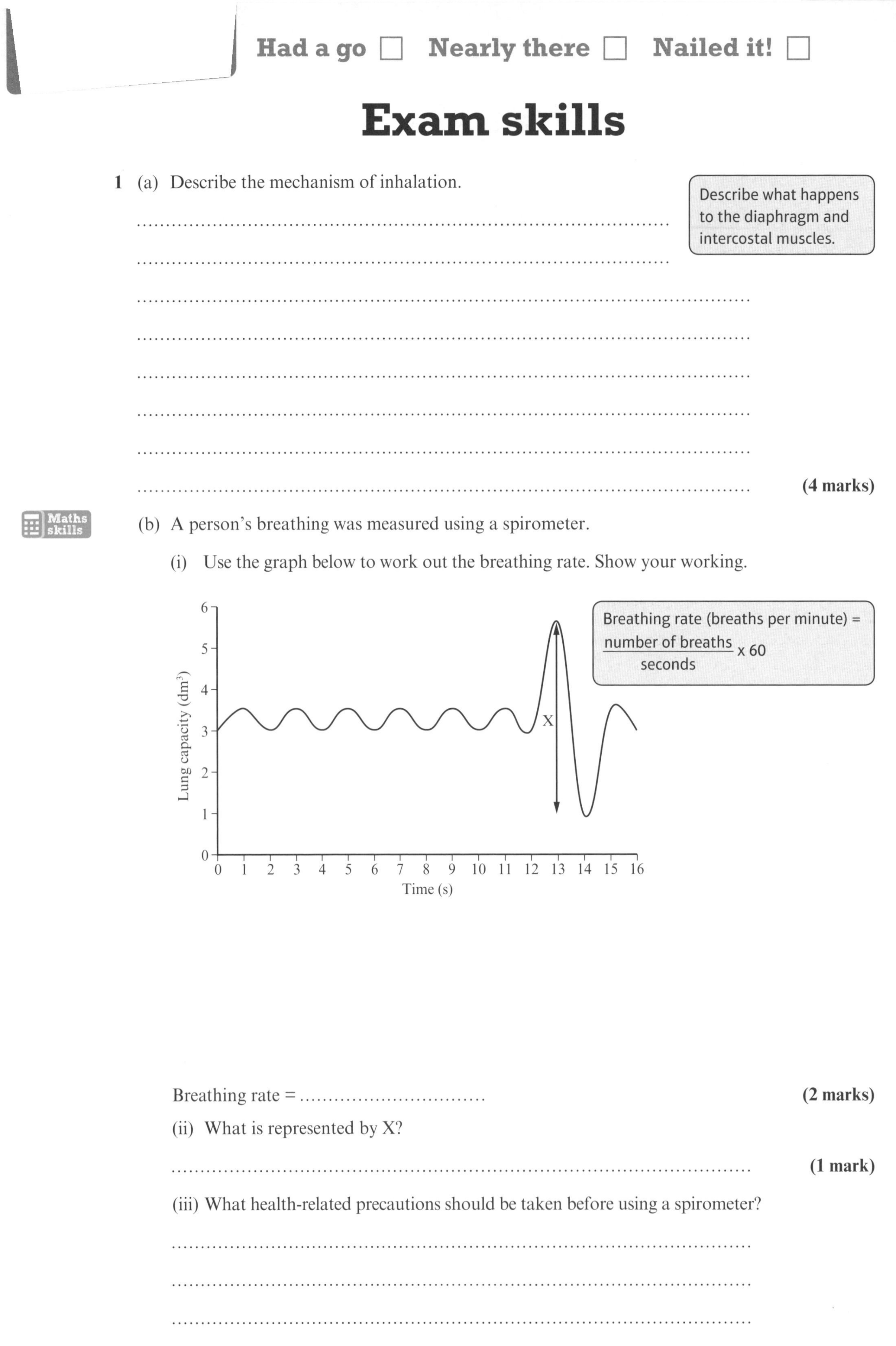

> Breathing rate (breaths per minute) = $\frac{\text{number of breaths}}{\text{seconds}} \times 60$

Breathing rate = ... **(2 marks)**

(ii) What is represented by X?

.... **(1 mark)**

(iii) What health-related precautions should be taken before using a spirometer?

....

....

....

.... **(2 marks)**

Types of pathogen

1 (a) What is a pathogen?

....

.... **(1 mark)**

(b) Which type of pathogen can only survive inside a host cell?

....

.... **(1 mark)**

(c) To which kingdom do pathogens with chitin walls belong?

....

.... **(1 mark)**

(d) Which **two** kingdoms contain eukaryotic pathogens?

....

....

....

.... **(2 marks)**

2 Which of the following statements are true?

Statement 1: Tuberculosis is caused by a bacterium.

Statement 2: Bacteria are much larger than viruses.

Statement 3: Malaria is caused by a virus.

☐ **A** 1 and 2 only

☐ **B** 2 and 3 only

☐ **C** 1 and 3 only

☐ **D** all of the above **(1 mark)**

Guided **3** Complete the table below to show which disease is caused by which type of pathogen.

Disease	Type of pathogen
influenza	virus
ring rot	....
tomato blight	....
ringworm	....

(4 marks)

Transmission of pathogens

1 (a) What is **direct transmission**?

....

.... **(1 mark)**

(b) Give an example of a disease that is spread by direct transmission.

....

.... **(1 mark)**

(c) Describe how plant fungal pathogens can be spread by direct transmission.

Think about how spores travel.

....

....

....

.... **(2 marks)**

2 Which of the following diseases is transmitted by indirect transmission?

☐ **A** malaria

☐ **B** HIV

☐ **C** cholera

☐ **D** anthrax **(1 mark)**

Guided

3 A farmer protects his crops from a fungal disease by spraying them with insecticide.

(a) Suggest how this could protect the plant from fungal disease.

Fungi on the infected ....

....

An insect eating the plant ....

....

Insecticide kills the insects ....

.... **(3 marks)**

(b) What else could the farmer spray the crops with to protect them from this fungal disease?

....

.... **(1 mark)**

Plant defences against pathogens

1 Which of the following plant primary defences prevents pathogens from entering the plant through the phloem?

☐ **A** stomatal closing

☐ **B** tylose formation

☐ **C** production of chemicals

☐ **D** callose deposition **(1 mark)**

2 Tylose formation protects a plant from infection.

(a) Describe how tylose formation protects the plant from infection.

Include plant chemicals in your answer.

....

....

....

....

....

.... **(3 marks)**

(b) Suggest a pathogen that tylose formation would protect against.

....

.... **(1 mark)**

Guided

3 Compare and contrast the primary defences of plants and animals.

Both have a permanent physical barrier against infection:

....

....

Both prevent entry of pathogens:

....

....

Both block movement of pathogens:

....

....

Both use chemicals:

....

.... **(4 marks)**

Animal defences against pathogens

1 (a) What is meant by a **non-specific defence**?

....

.... **(1 mark)**

(b) Which chemicals produced in your body are used to kill pathogens?

Think about chemicals found in your stomach, eyes and mouth.

....

....

....

.... **(2 marks)**

2 What are **commensal bacteria**?

☐ **A** bacteria that cause sickness and diarrhoea

☐ **B** non-harmful bacteria that live in the stomach and help to kill harmful bacteria

☐ **C** non-harmful bacteria that live on the skin and prevent harmful bacteria from growing there

☐ **D** harmful bacteria that live on the skin but are not harmful if they enter the body **(1 mark)**

3 Scabs form when the skin is cut.

(a) Describe how a scab is formed.

For 'Describe questions', you do not need to explain how or why, you only need to describe what is requested.

....

....

....

....

....

.... **(3 marks)**

Guided

(b) Explain the importance of mast cells at the site of the cut.

Mast cells produce the chemical histamine, which causes the blood capillaries surrounding the cut to vasodilate.

....

....

....

.... **(2 marks)**

Phagocytes

1 (a) What are the **two** main types of phagocyte?

....

.... **(2 marks)**

(b) Describe the mode of action of phagocytes.

....

....

....

....

....

.... **(3 marks)**

(c) Explain the role of cytokines in the non-specific immune response.

Cytokines are chemicals that attract phagocytes.

....

....

....

.... **(2 marks)**

Guided

2 Describe how an antigen-presenting cell is formed.

When a type of phagocyte called a .... engulfs a pathogen, it can display the pathogen's ....

....

....

....

....

....

....

.... **(4 marks)**

3 What are **opsonins**?

☐ **A** cells that can engulf pathogens

☐ **B** molecules that enhance phagocytosis

☐ **C** cells that can carry out exocytosis

☐ **D** molecules that activate the specific immune response **(1 mark)**

Lymphocytes

1 What is the role of a plasma cell?

☐ **A** to kill any cell infected with a virus

☐ **B** to secrete interleukins

☐ **C** to act as memory cells following an infection

☐ **D** to produce antibodies **(1 mark)**

2 (a) What is **clonal selection**?

....

....

....

.... **(2 marks)**

(b) Describe how T lymphocytes are activated.

Include antigen-presenting cells in your answer.

....

....

....

.... **(2 marks)**

(c) Name **four** types of T lymphocyte.

....

....

....

.... **(4 marks)**

(d) Explain the role of T lymphocytes in phagocytosis.

....

....

....

.... **(2 marks)**

Had a go ☐ Nearly there ☐ Nailed it! ☐

Immune responses

1 (a) Describe the structure of an antibody, relating the structure to its function.

Include constant and variable regions in your answer.

....

....

....

....

....

....

....

.... **(4 marks)**

(b) What is the role of antibodies in phagocytosis?

....

....

....

.... **(2 marks)**

2 The graph shows the primary and secondary immune response to infections.

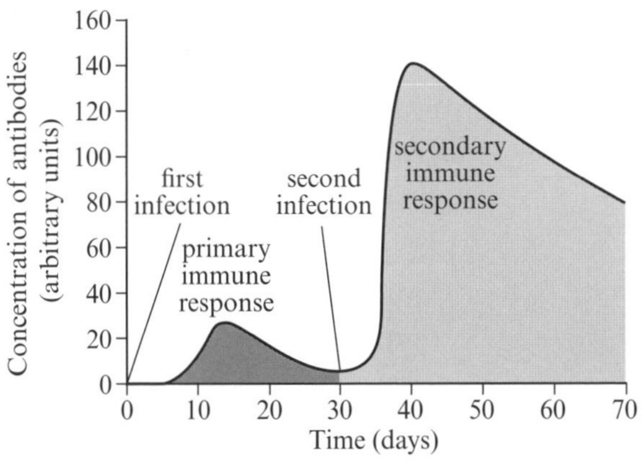

The graph shows the response to infections, not to injections.

Concentration of antibodies (arbitrary units)
160
140
120
100
80
60
40
20
0
first infection
primary immune response
second infection
secondary immune response
0 10 20 30 40 50 60 70
Time (days)

(a) What type of cell is responsible for the rapid secondary response?

.... **(1 mark)**

(b) What type of immunity is being shown by the secondary response here?

....

.... **(1 mark)**

Guided

(c) Describe how a secondary response can be caused artificially.

The body is injected with ....

....

The complementary lymphocytes are activated by the ....

.... **(2 marks)**

Types of immunity

1 What type of immunity is shown by antibodies passing to a baby through the mother's milk?

☐ **A** active natural immunity

☐ **B** passive natural immunity

☐ **C** active artificial immunity

☐ **D** passive artificial immunity

(1 mark)

2 (a) What is an **autoimmune disease**?

The answer needs to include a reference to the antigens found on your own body cells.

....

....

....

....

....

.... **(3 marks)**

(b) Give **two** examples of an autoimmune disease.

....

.... **(2 marks)**

3 If the body is cut with glass or metal, there is a risk of infection by bacteria that cause tetanus. This question is about a person who has not had a tetanus vaccination and has been infected by the tetanus pathogen.

(a) Explain why the treatment given involves injecting antibodies to tetanus, rather than injecting tetanus antigens.

This 'explain' question requires you to set out reasons for injecting antibodies using your knowledge of the immune system.

....

....

....

....

....

.... **(3 marks)**

(b) What type of immunity would an injection of tetanus antibodies be?

.... **(1 mark)**

The principles of vaccination

1 Which of the following statements are true?

Statement 1: Vaccination is the injection of antibodies into the body.

Statement 2: Vaccination is the injection of live, harmless virus into the body.

Statement 3: Vaccination is the injection of dead virus into the body.

Statement 4: Vaccination is the injection of antigens into the body.

☐ **A** 1, 2 and 3 only

☐ **B** 2, 3 and 4 only

☐ **C** 1, 3 and 4 only

☐ **D** all of the above **(1 mark)**

2 (a) Explain what is meant by **herd immunity**.

....

....

....

....

....

.... **(3 marks)**

(b) Explain what may happen if less than 90 per cent of the population are vaccinated against a particular disease.

Include in your answer the individuals who are not vaccinated.

....

....

....

.... **(2 marks)**

3 List **three** sources of antigens when making a vaccine.

....

....

....

....

....

.... **(3 marks)**

Had a go ☐ Nearly there ☐ Nailed it! ☐

Vaccination programmes

1 What is ring vaccination?

☐ **A** When everyone in a population is vaccinated

☐ **B** When most people in a population are vaccinated

☐ **C** When all of the population in an area where a disease has broken out are vaccinated

☐ **D** When most of the population in an area where a disease has broken out are vaccinated

(1 mark)

2 (a) Define the term **epidemic**.

Specify the meaning of the word or term.

....

....

(1 mark)

(b) Describe how vaccines are used to prevent epidemics.

....

....

....

....

(2 marks)

3 New influenza vaccines are produced every year and given to people in at-risk groups.

(a) Give **three** examples of at-risk groups who would need vaccination.

....

....

....

(3 marks)

(b) Explain why a new influenza vaccine is needed every year.

Include the idea of mutations in your answer.

....

....

....

....

(2 marks)

Antibiotics

1 The graphs show changes in the quantities of an antibiotic used in a hospital and the percentage of infections caused by bacteria resistant to the antibiotic over the same time period.

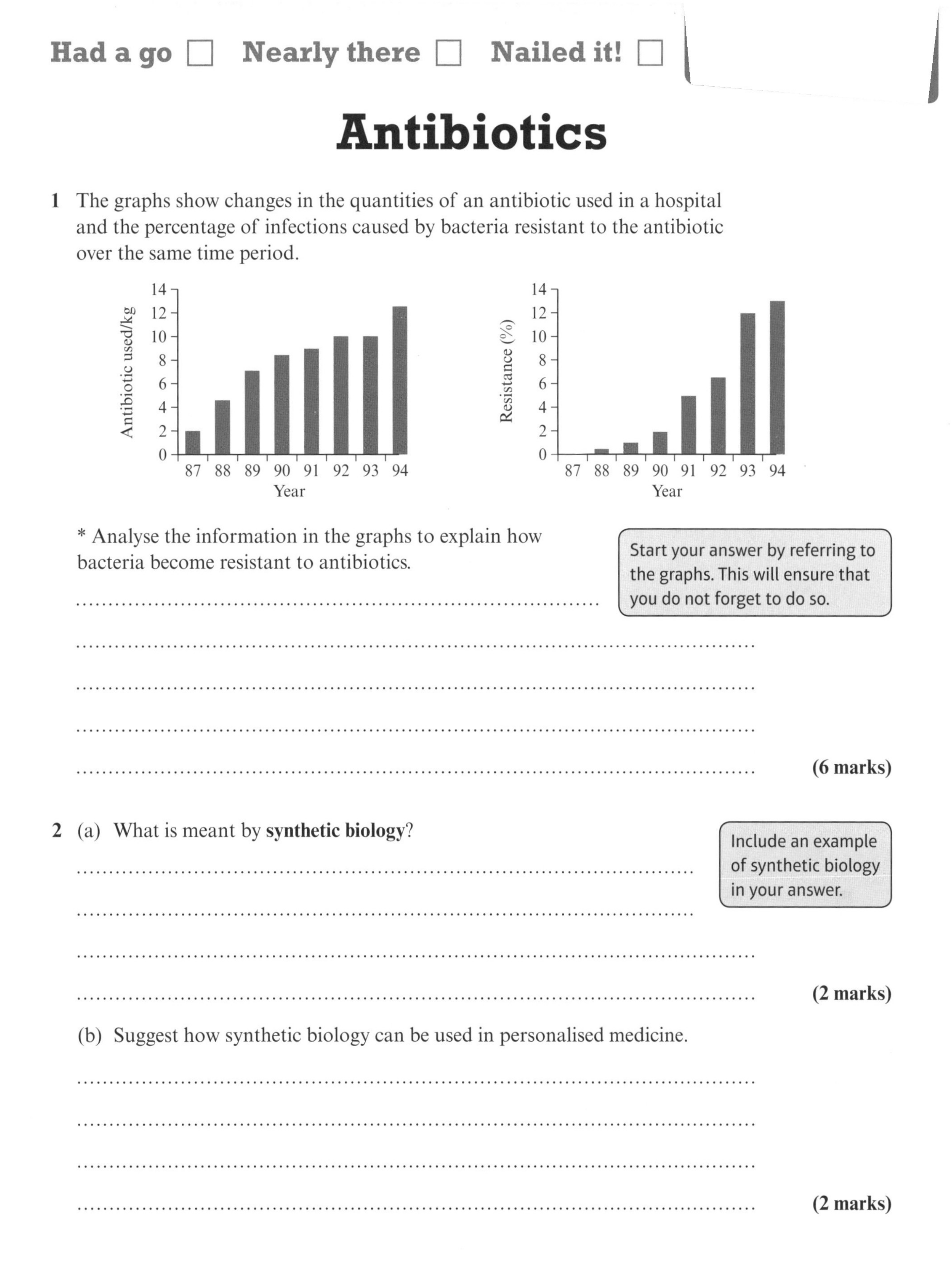

* Analyse the information in the graphs to explain how bacteria become resistant to antibiotics.

Start your answer by referring to the graphs. This will ensure that you do not forget to do so.

....

....

....

....

.... **(6 marks)**

2 (a) What is meant by **synthetic biology**?

Include an example of synthetic biology in your answer.

....

....

....

.... **(2 marks)**

(b) Suggest how synthetic biology can be used in personalised medicine.

....

....

....

.... **(2 marks)**

Exam skills

1 Tuberculosis affects 2 billion people worldwide.

(a) Name the pathogen that causes tuberculosis.

.... **(1 mark)**

(b) Describe how tuberculosis can be spread from person to person.

....

....

....

.... **(2 marks)**

(c) What type of drug can be used to treat tuberculosis?

.... **(1 mark)**

(d) A vaccine has been developed to protect against tuberculosis. Explain how the vaccine works.

Discuss the role of antigens and the response of the lymphocytes.

....

....

....

....

....

....

....

....

....

.... **(5 marks)**

(e) Suggest why people in the UK are no longer routinely vaccinated against tuberculosis.

....

.... **(1 mark)**

Practical skills

Measuring biodiversity

1 (a) What is the difference between species evenness and species richness?

....

....

....

.... **(2 marks)**

(b) In the study of a field habitat, describe how the species evenness could be measured.

Think about how you could use quadrats.

....

....

....

.... **(2 marks)**

2 Which of the following is an example of random sampling?

☐ **A** taking opportunistic samples

☐ **B** using a number generator to produce coordinates to place quadrats

☐ **C** choosing two different areas in a habitat to sample

☐ **D** taking samples at fixed intervals across a habitat **(1 mark)**

Guided

3 * With reference to sampling a large forest, analyse the advantages and disadvantages of non-random sampling.

Discuss how the advantages and disadvantages affect the data collection.

....

....

....

....

....

....

....

....

.... **(6 marks)**

Practical skills

Sampling methods

1 Describe how to use a random frame quadrat.

....

....

....

.... **(2 marks)**

2 Which of the following are collection methods for animals?

Method 1: kick sampling

Method 2: sweep net

Method 3: pitfall trap

Method 4: belt transect

☐ **A** 1, 2 and 3 only

☐ **B** 2, 3 and 4 only

☐ **C** 1, 3 and 4 only

☐ **D** all of the above **(1 mark)**

3 A belt transect is laid along a seashore and the plant species are measured at intervals along it.

(a) What type of sampling is this?

.... **(1 mark)**

(b) What information does a belt transect give you that random quadrats alone do not?

A belt transect measures changes across an area.

....

....

....

.... **(2 marks)**

Simpson's index of diversity

1 What does Simpson's index of diversity measure?

☐ **A** the level of genetic biodiversity of an area

☐ **B** the level of biodiversity of a habitat

☐ **C** the total number of species in an area

☐ **D** the total number of individuals in a habitat **(1 mark)**

Maths skills

2 The data below have been collected from the same aquatic habitat over two different years, and the diversity calculated using the formula $D = 1 - \left[\Sigma\left(\frac{n}{N}\right)^2\right]$.

n = The number of individuals of one species in the habitat
N = The total number of individuals of all species in the habitat

	2010			2015		
Species	**Number of individuals**	$\frac{n}{N}$	$\left(\frac{n}{N}\right)^2$	**Number of individuals**	$\frac{n}{N}$	$\left(\frac{n}{N}\right)^2$
water snail	1	0.063	0.004	6	0.300	0.090
freshwater shrimp	2	0.125	....	4	0.200	0.040
mayfly nymph	1	0.063	0.004	7	0.350	0.123
bloodworm	7	0.438	0.192	2	0.100	....
rat-tailed maggot	5	0.313	0.098	1	0.050	0.003
	N = ...	Σ =	0.314	N = 20	Σ =	0.266
		$D = 1 - 0.314$			$D = 1 - 0.266$	
		D =	**0.686**		D =	....

(a) Complete the table. **(4 marks)**

(b) Describe how the biodiversity of the area changed between 2010 and 2015.

....

....

....

....

.... **(3 marks)**

(c) Suggest what could have caused this change.

Here you need to draw on your biological knowledge to answer this question.

....

....

....

.... **(2 marks)**

Factors affecting biodiversity

1 (a) What is **genetic diversity**?

....

.... **(1 mark)**

(b) Why might a population have low genetic diversity?

....

.... **(1 mark)**

Guided

(c) Describe how genetic diversity can be calculated.

By working out the number of genes that are polymorphic.

....

.... **(1 mark)**

2 What does **polymorphic** mean?

☐ **A** a gene with two or more alleles

☐ **B** a gene with only one allele

☐ **C** the inheritance of two identical alleles

☐ **D** the inheritance of two different alleles **(1 mark)**

3 (a) Give **two** examples of human activities that reduce genetic diversity.

....

....

....

.... **(2 marks)**

Maths skills

(b) Work out the genetic diversity of a population with 300 out of 2800 polymorphic gene loci.

$$\text{Proportion of polymorphic gene loci} = \frac{\text{number of polymorphic gene loci}}{\text{total number of loci}}$$

.... **(3 marks)**

(c) Explain what this proportion of polymorphic gene loci means.

....

.... **(1 mark)**

Maintaining biodiversity

1 (a) What is **in-situ conservation**?

....

.... **(1 mark)**

(b) List **two** advantages of in-situ conservation.

Think about the genetic diversity of a population.

....

....

....

.... **(2 marks)**

2 Which of the following are reasons for maintaining biodiversity?

Reason 1: to protect species from extinction

Reason 2: to enjoy the natural beauty of the habitat

Reason 3: to allow pollination of plants by insects

Reason 4: ecotourism

☐ **A** 1, 2 and 3

☐ **B** 1, 2 and 4

☐ **C** 2, 3 and 4

☐ **D** all of the above **(1 mark)**

Guided

3 (a) Describe how conservation agreements have helped to maintain global and local biodiversity.

The Convention on International Trade in Endangered Species (CITES) prevents the trade of endangered wild plants, animals and animal parts.

....

....

....

.... **(2 marks)**

(b) Explain how a ban on the trade of endangered animals increases the genetic variation of that species.

....

....

....

.... **(2 marks)**

Classification

1 W, X, Y and Z are related species. Y and Z are evolved from species X. Species W is the least related to the others. Which of the phylogenetic trees below represents these relationships?

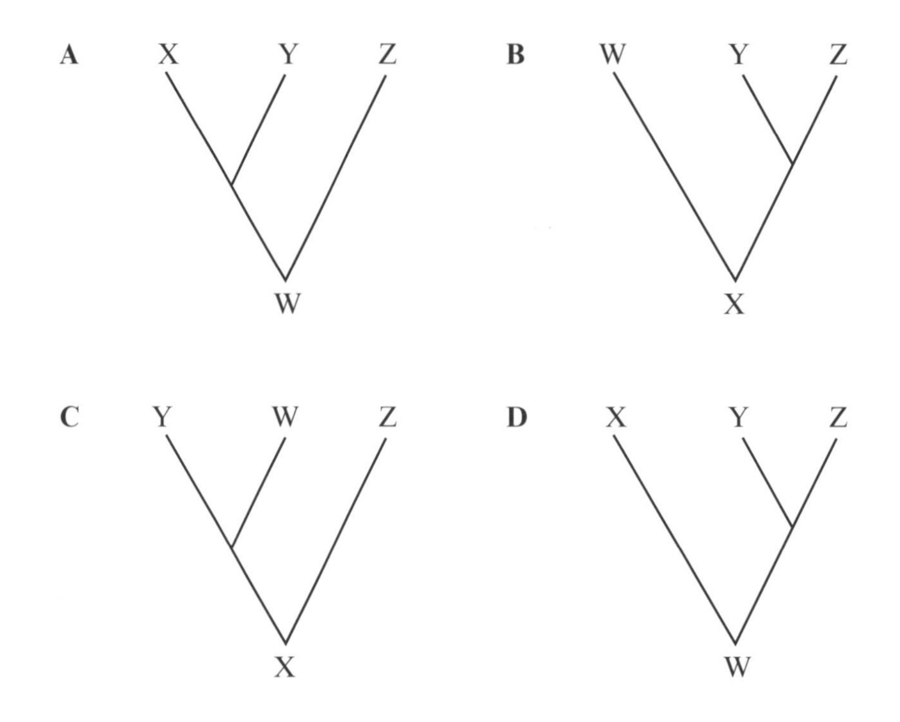

Your answer .... **(1 mark)**

2 Complete the classification hierarchy table for the wolf, *Canis lupus*.

Domain	Kingdom	....	Class	Order	....	Genus	Species
Eukarya	Animalia	Chordata	Mammalia	Carnivora	Canidae	....	....

(4 marks)

3 Explain how biological molecules could be used to examine how closely species are related.

....

....

....

....

....

....

....

.... **(4 marks)**

The five kingdoms

1 Which of the following descriptions best describes an organism from the Protoctista kingdom?

☐ **A** eukaryotic, multicellular and autotrophic

☐ **B** prokaryotic, single-celled and heterotrophic

☐ **C** eukaryotic, single-celled and autotrophic

☐ **D** eukaryotic, multicellular and heterotrophic **(1 mark)**

2 (a) What does **autotrophic** mean?

....

.... **(1 mark)**

(b) What types of organisms are autotrophic?

....

....

....

.... **(2 marks)**

Guided

(c) List the methods used to organise species into taxa.

Observation of features, such as anatomy

....

....

....

....

.... **(3 marks)**

3 (a) What are the three domains?

....

.... **(1 mark)**

(b) Give **one** similarity and **one** difference between the organisms in the Eubacteria and Archaea domains.

Think about the cell structure of the organisms in each domain.

....

....

....

.... **(2 marks)**

Types of variation

1 (a) What is **continuous variation**?

....

.... **(1 mark)**

(b) Give an example of continuous variation.

....

.... **(1 mark)**

(c) Describe the causes of continuous variation.

....

....

....

.... **(2 marks)**

2 Which graph best shows discontinuous variation?

A

B

C

D

Your answer .... **(1 mark)**

Maths skills

3 The reaction times of a group of students were recorded before and after drinking caffeinated cola. Student's t-test was applied to the data. The t-test gave a value of 0.389. Are the reaction times after drinking caffeine significantly quicker than before drinking caffeine?

The t-test value must be greater than the critical value at $p = 0.05$ for the data to be considered significant.

	Mean reaction time (s)
Before caffeine	0.323
After caffeine	0.276

	Significance level	
Degrees of freedom	$p = 0.05$	$p = 0.01$
1	12.7	63.7
2	4.3	9.9

....

....

....

....

....

.... **(3 marks)**

Evolution by natural selection

1 Explain why dolphins and sharks both have smooth, streamlined bodies, despite not being closely related.

> Use the words 'convergent', 'selective' and 'adaptations' in your answer.

....

....

....

....

....

.... **(3 marks)**

2 (a) The Chilean dolphin (*Cephalorhynchus eutropia*) and Hector's dolphin (*Cephalorhynchus hectori*) look very alike. Explain why they are two distinct species.

....

....

....

.... **(2 marks)**

(b) What genus do these dolphins belong to?

.... **(1 mark)**

(c) *Explain how natural selection might have given rise to the adaptations shown by the Chilean dolphin.

> Aim to give a detailed explanation using the information given and your own knowledge. Give evidence to support your points. Make sure your answer is clear and logically structured and that your points are linked.

....

....

....

....

....

....

....

....

.... **(6 marks)**

Practical skills

Evidence for evolution

1 A researcher concluded that wild horses (*Equus ferus*) evolved from an extinct species, *Plesippus*.

(a) What evidence could the researcher use to validate this conclusion?

....

....

....

....

....

.... **(3 marks)**

(b) Describe some limitations of collecting evidence for evolution.

Think about parts of a body that decay.

....

....

....

....

....

....

....

.... **(4 marks)**

2 Use the DNA sequences below to show which of the following species are more closely related.

Species 1: ATTGCGATACCGGCG

Species 2: AATCGCATAGCGCGC

Species 3: AATCGGATAGCGCGG

Species 4: AATCCGATAGCGCGG

☐ **A** 1 and 2

☐ **B** 2 and 3

☐ **C** 3 and 4

☐ **D** 1 and 3 **(1 mark)**

Exam skills

Practical skills

1 A study was carried out to measure the biodiversity of a habitat.

(a) What is meant by **species richness**?

....

.... **(1 mark)**

Practical skills

(b) Describe the best method for measuring the abundance of plants in two areas of the habitat.

....

....

....

.... **(2 marks)**

(c) Simpson's index of diversity was calculated for one area of the habitat and gave a score of 0.89. Explain what this means.

The closer the index of diversity score is to 1, the higher the level of biodiversity.

....

....

....

.... **(2 marks)**

(d) Ten years ago, the index of diversity score for the same area was 0.56. Suggest what might have caused the score to change.

....

....

....

.... **(2 marks)**

(e) Explain how human activity can reduce the level of biodiversity.

....

....

....

.... **(2 marks)**

The need for communication

1 What are the two major communication systems within the human body?

....

.... **(2 marks)**

2 (a) Define the term **stimulus**.

Use the words 'change' and 'behaviour' in your definitions.

....

.... **(1 mark)**

(b) Define the term **response**.

....

.... **(1 mark)**

3 Which factor does **not** affect the way enzymes function in cells?

☐ **A** light intensity

☐ **B** pH

☐ **C** temperature

☐ **D** water balance **(1 mark)**

4 Explain the role of tissue fluid in maintaining a suitable internal environment.

Consider the substances that move into the tissue fluid from the blood and out of cells into tissue fluid.

....

....

....

....

....

.... **(3 marks)**

5 A good communication system is required to ensure different parts of the body work together. Give **three** features of a good communication system.

....

....

....

....

....

.... **(3 marks)**

Principles of homeostasis

1 Are the following primarily controlled by the nervous system or the hormonal system?

	Nervous	Hormonal
body temperature	....	....
blood glucose concentration	....	....
water potential of the blood	....	....
carbon dioxide concentration of the blood	....	....

(4 marks)

2 Define the term **negative feedback**.

(2 marks)

3 Explain the effect of increasing core body temperature on the function of enzymes. You may use a diagram to support your answer.

When asked to draw a diagram, make sure you include clear and accurate labels and use a ruler where necessary. If drawing a graph, make sure you also label axes with the name/variable and units. An annotated diagram similar to this would be useful.

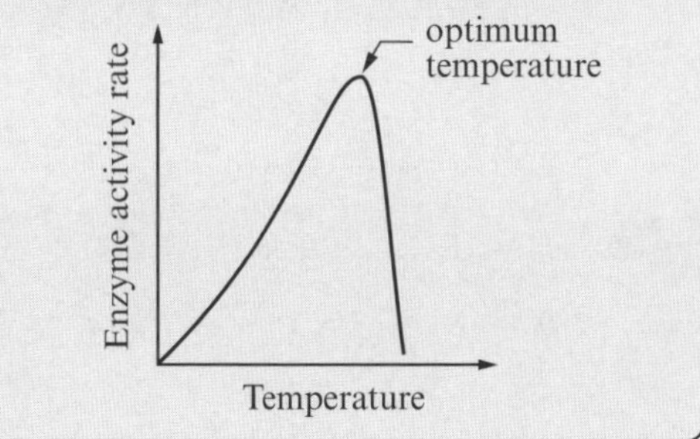

As the temperate increases, so does the ....

....

....

....

....

....

.... **(4 marks)**

Temperature control in endotherms

1 How do endotherms maintain their body temperature?

....

....

....

.... **(2 marks)**

2 Which one of the following statements about endotherms is false?

☐ **A** Endotherms can inhabit the colder parts of the planet.

☐ **B** Panting is a mechanism used by endotherms to cool down.

☐ **C** The arterioles under the skin undergo vasodilation if the core temperature becomes too high.

☐ **D** The body temperature of endotherms varies with the external environment. **(1 mark)**

Guided

3 *Explain how endotherms use negative feedback to maintain body temperature.

Make sure you structure your answer logically, showing how the points you make are related or follow on from each other where appropriate. You should also support your points with relevant biological facts and/or evidence.

A fall in the environmental temperature is first detected by ....

....

....

....

....

....

....

....

....

.... **(6 marks)**

4 Explain why vasodilation must take place in the arterioles and not the capillaries that lie near the surface of the skin.

....

....

....

.... **(2 marks)**

Temperature control in ectotherms

1 Which one of the following statements about ectotherms is false?

☐ **A** Ectotherms can live in a wider variety of climates than endotherms.

☐ **B** Ectotherms convert more of what they eat into growth.

☐ **C** Ectotherms require less food than endotherms.

☐ **D** Increasing their bodily surface area is one way ectotherms get warm. **(1 mark)**

2 Describe the type of behaviour you would expect an ectotherm to exhibit on a cool, bright, sunny morning to increase its core body temperature.

....

....

....

.... **(2 marks)**

3 Give **two** advantages of being an ectotherm.

....

....

....

.... **(2 marks)**

4 Suggest why cellular activity will decrease if the core body temperature of an ectotherm gets too high.

Consider what controls reactions inside cells.

....

....

....

....

....

.... **(3 marks)**

5 How is it possible for some ectotherms to go for weeks without feeding?

....

.... **(1 mark)**

Excretion

1 Suggest what is meant by the term **excretion**.

....

....

....

.... **(2 marks)**

2 (a) Name the two excretory products that are produced in relatively large quantities in the body.

One is a gas and one is a nitrogenous compound.

....

.... **(2 marks)**

(b) Explain the link between an excessive consumption of protein and the generation of a high concentration of nitrogenous excretory products.

....

....

....

.... **(2 marks)**

Guided

3 Explain how respiration can cause a fall in blood pH.

Consider using equations in your answer. This is an 'Explain' question, so it is not enough to state your point - you must also give reasons or say why you have come to that conclusion.

Respiration produces carbon dioxide. ....

....

....

....

....

....

....

.... **(3 marks)**

4 Which one of the following statements about excretion is true?

☐ **A** Ammonia is a safer substance to transport in the blood than urea.

☐ **B** Deamination is the removal of an amine group from an amino acid.

☐ **C** Egestion and excretion are the same thing.

☐ **D** Urea diffuses into the lungs and is breathed out. **(1 mark)**

The liver – structure and function

1 Give the names of the two blood vessels which deliver blood to the liver.

....

.... **(2 marks)**

2 (a) State the differences between the blood in the hepatic portal vein and the blood in the hepatic vein.

Give a concise answer, there is no need for explanation.

....

....

....

....

....

.... **(3 marks)**

(b) To what organ does the bile duct lead?

.... **(1 mark)**

(c) State the role of Kupffer cells.

....

.... **(1 mark)**

3 Which one of the following is **not** a role of the liver?

☐ **A** pumping blood

☐ **B** storing carbohydrates

☐ **C** storing vitamins

☐ **D** synthesising proteins **(1 mark)**

4 Liver cells, or hepatocytes, are highly metabolically active. They carry out many processes which require energy. Which organelle would you expect to find in hepatocytes in large numbers and why?

....

....

....

.... **(2 marks)**

Had a go ☐ Nearly there ☐ Nailed it! ☐

The kidney – structure and function

1 In the space below, draw a cross-section of a kidney and label the cortex, medulla and pelvis.

Do not worry about drawings being perfect. Here it is the boundaries between the tissues that are important.

(3 marks)

2 In which region would you expect to find the proximal convoluted tubules of the nephron?

.... **(1 mark)**

3 Which **two** structures are involved in the process of ultrafiltration?

....

.... **(2 marks)**

4 Which sequence shows the correct order of structures in the excretory system through which urine passes?

☐ **A** Bowman'scapsule, PCT, DCT, loop of Henle, glomerulus, collecting duct

☐ **B** glomerulus, Bowman'scapsule, PCT, loop of Henle, DCT, collecting duct

☐ **C** loop of Henle, PCT, glomerulus, DCT, Bowman'scapsule, collecting duct

☐ **D** Bowman'scapsule, glomerulus, DCT, loop of Henle, PCT, collecting duct **(1 mark)**

5 Describe the main differences in the contents of blood arriving in the kidney through the renal artery and the blood leaving through the renal vein.

....

....

....

....

....

....

.... **(3 marks)**

Osmoregulation

1 Explain how the glomerulus generates hydrostatic pressure.

....

....

....

.... **(2 marks)**

Guided

2 Complete the table below to show what you would normally expect to find in blood in the renal artery, in the glomerular filtrate and in the urine of a healthy person.

The glomerular filtrate is the name given to the fluid inside the nephron.

Substance	Blood	Glomerular filtrate	Urine
red blood cells	✓	✗	✗
amino acids			
glucose			
urea			

(3 marks)

3 Explain how the difference in content between the glomerular filtrate and the urine is achieved.

You need to explain what has happened in the nephron to bring about the changes in levels of amino acids, glucose, urea and water.

....

....

....

....

....

....

....

.... **(4 marks)**

4 The tissue fluid of the medulla has a lower water potential than the tissue fluid of the cortex. Explain what is responsible for this difference.

....

....

....

.... **(2 marks)**

Kidney failure and urine testing

1 Renal dialysis is one option for a patient with kidney failure. Outline what is meant by the term **dialysis**.

....

....

....

....

....

.... **(3 marks)**

2 Following dialysis do you expect the blood concentrations of the following substances to go up, down or stay the same?

	Go up	Go down	Stay the same
urea			
glucose			
amino acids			
salts			

(4 marks)

3 *Explain how a pregnancy test strip might work.

Answer this in five points: (1) which hormone you will test for; (2) what bodily liquid you will test; (3) If the hormone is present in the urine, what will it bind to on the test strip?; (4) What are the fixed antibodies on the test strip specific to?; (5) What will you see if the test is positive? and (6) Explain why you would see this.

....

....

....

....

....

....

....

....

....

....

....

....

.... **(6 marks)**

Exam skills

1 Explain how increased sweating is involved in the regulation of body temperature.

It is not warm sweat that leaves the body but water vapour which has been created using energy from the blood.

....

....

....

....

....

.... **(3 marks)**

2 This image of a human head and neck shows part of the CNS.

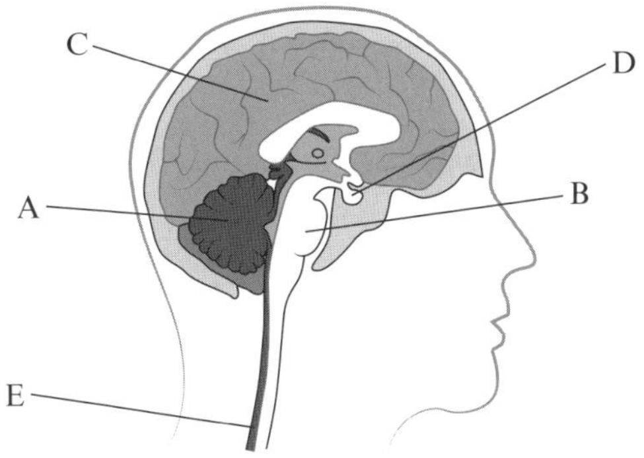

Give the letter and name of the structure involved in thermoregulation.

....

....

(2 marks)

Maths skills

3 Research has been carried out on the effect of amphetamines on sporting performance. The table below shows data from a study of volunteers.

Event	Average performance (s)		
	After taking placebo	After taking amphetamines	Improvement (%)
200 m freestyle	136.88	135.94	0.69
200 m backstroke	159.80	158.32	0.93
200 m breaststroke	171.87	170.22	

(a) Using the data, calculate the percentage improvement for breaststroke.

Always do a rough 'guesstimate' first with percentage calculations.

....

.... **(1 mark)**

(b) Discuss whether the use of amphetamines as performance-enhancing drugs should be made legal.

In 'discuss' questions, you need to put both sides of an argument.

....

....

....

....

....

.... **(3 marks)**

Sensory receptors

Guided

1 *A transducer converts energy from one form into another. Outline the processes that take place in a Pacinian corpuscle when mechanical pressure is converted into a nervous impulse.

The stretch mediated sodium ion channels are too narrow to allow sodium ions to pass through in its resting state. When pressure ....

....

....

....

....

.... **(6 marks)**

2 Give **two** places in the body where you would expect to find chemoreceptors.

....

.... **(2 marks)**

3 Which of the following statements about sensory receptors is false?

☐ **A** All sensory receptors can act as transducers.

☐ **B** Na^+ channels play an important role in sensory receptors.

☐ **C** Rods and cones are examples of photoreceptors.

☐ **D** Sensory receptors can be found at the end of motor neurones. **(1 mark)**

4 What type of stimuli are detected by **mechanoreceptors**?

....

....

....

.... **(2 marks)**

5 What type of energy are different stimuli converted into by sensory receptors?

....

.... **(1 mark)**

Types of neurone

1 The figure represents a cross-section through a myelinated neurone.

(a) Identify structures X, Y and Z.

X ....

Y ....

Z .... **(3 marks)**

(b) What is the gap between neighbouring Schwann cells called?

.... **(1 mark)**

(c) What effect does myelination have on a nervous impulse?

....

.... **(1 mark)**

Guided

2 Which of the following statements relate to sensory, motor and/or relay neurones?

	Sensory	Motor	Relay
has a cell body in the CNS but an axon in the peripheral nervous system (PNS)			
whole cell found in the CNS			
carries an action potential from the CNS to an effector organ			
has a large cell body at one end of the cell		✓	✓
has a cell body to one side of the cell which is **not** in the CNS			
has the longest axon			
is most likely to form the largest number of synapses with other neurones			

(6 marks)

3 Grey matter is so called because it has a light grey appearance. It primarily contains cell bodies, dendrites and unmyelinated neurones. Suggest what might be the consequences of all nervous tissue being 'grey matter'?

....

....

....

.... **(2 marks)**

Had a go ☐ Nearly there ☐ Nailed it! ☐

Action potentials and impulse transmission

1 Put the following into the correct sequence to represent the generation of an action potential.

A Voltage-gated Na^+ channels close.

B The threshold potential of about –40 mV is reached.

C Some gated Na^+ channels are stimulated to open.

D Na^+ ions flood into the cell, making the inside positive relative to the outside.

E Voltage-gated K^+ channels open and K^+ diffuses out of the cell.

F The potential difference across the cell membrane reaches +40 mV.

G The potential difference falls below the resting potential.

.... **(3 marks)**

2 Action potentials are always of the same intensity. Explain how stimuli of different strengths can be detected.

....

....

.... **(2 marks)**

3 Explain how it is possible to maintain a resting potential at about −60 mV.

> The two key proteins involved in maintaining a resting potential are the Na^+/K^+ pump and K^+ channels.

....

....

....

....

....

.... **(4 marks)**

4 The image below left represents a cross-section of an axon at rest. Draw a similar diagram to show how the axon might appear halfway through an action potential.

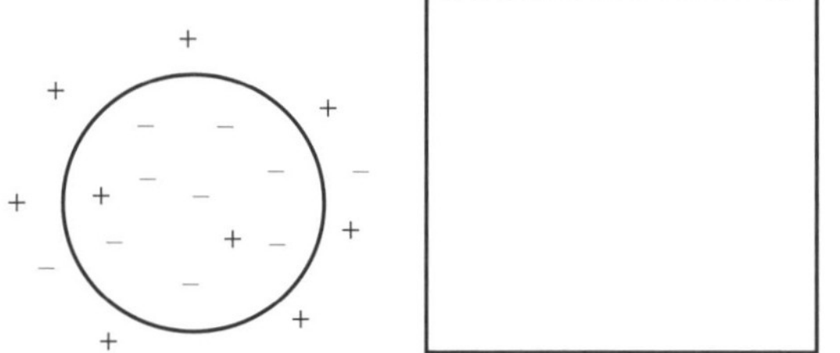

(2 marks)

Structure and roles of synapses

1 What are neurotransmitters and where do they act?

....

....

....

....

....

.... **(3 marks)**

2 The following events occur at a synapse. Put them in the correct order.

.... **A** Acetylcholine binds to receptors on Na^+ ion channels in the post-synaptic membrane.

.... **B** Acetylcholine diffuses across the synaptic cleft.

.... **C** Na^+ ion channels open on the post-synaptic membrane.

.... **D** Vesicles of acetylcholine fuse with the pre-synaptic membrane and acetylcholine is released. **(2 marks)**

3 Explain the importance of acetylcholinesterase at the neuromuscular junction.

First outline what acetylcholine does. Then explain what would happen if it was inhibited or absent.

....

....

....

....

....

.... **(3 marks)**

4 There are two types of summation. Name them and summarise the differences between them.

....

....

....

....

....

.... **(3 marks)**

Endocrine communications

1 Which of the following is **not** an endocrine gland or **does not** contain endocrine tissue?

☐ **A** gall bladder ☐ **C** pancreas

☐ **B** ovaries ☐ **D** pituitary **(1 mark)**

2 Outline the differences between exocrine glands and endocrine glands. Include examples of both in your answer.

....

....

....

....

....

.... **(4 marks)**

3 Antidiuretic hormone (ADH) is released from the pituitary gland in response to a lowering in the water potential of the blood.

(a) Where are the target cells for ADH?

.... **(1 mark)**

(b) How does ADH reach these target cells?

.... **(1 mark)**

4 Which one of the following statements about hormones is true?

☐ **A** All hormones are protein based.

☐ **B** All hormones are transported in the blood.

☐ **C** Hormones bind permanently to receptors.

☐ **D** No hormones can pass into the cytoplasm of cells. **(1 mark)**

5 Give **four** differences between hormonal communication and nervous communication.

....

....

....

....

....

....

.... **(4 marks)**

Endocrine tissues

1 Describe how adrenaline interacts with target cells.

....

.... **(1 mark)**

2 Adrenaline can be described as a first messenger molecule. It interacts with target cells and induces a response inside the cell. Describe the role of adenyl cyclase in this process.

Remember to talk about the conformational change that occurs in the adrenaline receptor in your answer.

....

....

....

....

....

.... **(3 marks)**

3 One of the actions of adrenaline is to bring about the dilation of pupils. Suggest how this might support the 'fight or flight' response in mammals?

....

....

....

.... **(2 marks)**

4 (a) In which part of the adrenal gland is adrenaline produced?

.... **(1 mark)**

(b) Besides adrenaline and noradrenaline, the adrenal glands produce three other types of hormone. Name **two** of these.

....

.... **(2 marks)**

5 The pancreas contains both exocrine and endocrine tissue.

(a) Describe how the endocrine tissue is arranged.

.... **(1 mark)**

(b) Name **one** substance produced by the exocrine tissue.

.... **(1 mark)**

Regulation of blood glucose

1 (a) Where is insulin produced?

.... **(1 mark)**

(b) Insulin release leads to a decrease in blood glucose concentration. How is this achieved?

....

....

....

....

....

.... **(3 marks)**

2 Which statement about blood glucose regulation is false?

☐ **A** A fall in blood glucose concentration is detected by liver cells.

☐ **B** Glucagon causes liver cells to break down glycogen into glucose.

☐ **C** Insulin targets liver and muscle cells.

☐ **D** Insulin causes more glucose to be converted to fats. **(1 mark)**

Guided

3 An increase in the blood glucose concentration results in insulin being produced by β-cells. Outline how high levels of glucose in β-cells bring about the release of insulin.

Glucose is metabolised to produce ATP, ...

....

....

....

....

....

....

....

....

.... **(5 marks)**

4 What is the clinical term that describes very low blood glucose levels?

.... **(1 mark)**

Diabetes mellitus

1 What are patients suffering from diabetes mellitus unable to do?

....

.... **(1 mark)**

2 A 55-year-old man visits the doctor complaining of tiredness and lethargy. According to his BMI he is borderline overweight/obese. The doctor suspects diabetes. Suggest two questions which the doctor might ask the man and one test she might carry out.

Here you need to apply your biological knowledge and understanding.

....

....

....

....

....

.... **(3 marks)**

3 Type I diabetes is characterised by an inability to produce insulin. It is thought this might be the result of an autoimmune response. Explain how an autoimmune response can stop insulin production.

....

....

....

.... **(2 marks)**

4 Stem cells show potential as a future treatment of diabetes mellitus. Which of the following statements about stem cells is incorrect?

☐ **A** Stem cells are made in the brain stem.

☐ **B** Stem cells are not fully differentiated.

☐ **C** Stem cells can differentiate to become specialised cells such as β-cells.

☐ **D** The placenta is a source of stem cells. **(1 mark)**

5 Healthcare professionals are concerned that the incidence of type II diabetes is likely to increase significantly in coming decades. Suggest why they are predicting this increase.

....

....

....

.... **(2 marks)**

Plant responses to the environment

1 Directional plant growth responses are called tropisms. Give **two** examples of tropisms.

....

.... **(2 marks)**

Guided

2 *A lack of water to the roots of a plant stimulates a stress response. Explain how this stimulation leads to a stress response and give the consequences to the plant.

Aim to give a detailed explanation. Give evidence to support your points. Make sure your answer is clear and logically structured and that your points are linked.

When levels of soil water fall and transpiration is under threat ....

....

....

....

....

....

....

.... **(6 marks)**

3 Name the plant hormone that initiates seed germination.

.... **(1 mark)**

4 Which one of the following stimuli **does not** bring about a directional growth response in plants?

☐ **A** gravity

☐ **B** oxygen levels

☐ **C** light

☐ **D** water

(1 mark)

5 When a seed germinates it does not obtain energy through photosynthesis. Explain where the energy for the initial growth of a seed comes from.

....

....

....

....

.... **(3 marks)**

6 What is a non-directional response in a plant called?

.... **(1 mark)**

Controlling plant growth

1 Describe what happens to the growth of a plant shoot if the apical tip is removed.

....

.... **(1 mark)**

2 (a) Name the hormone that reduces in concentration in the shoot when the apical tip is removed.

....

.... **(1 mark)**

(b) Experiments show that high levels of this hormone keep abscisic acid levels high in the bud. When the tip is removed, lateral buds start to grow. Use this information to explain how trimming hedge tops can help them to bush out.

....

....

....

.... **(2 marks)**

3 Which of the following is a true statement about gibberellins?

☐ **A** They control internodal growth.

☐ **B** They are present in high concentrations in dwarf pea plants.

☐ **C** In high concentrations, they prevent flowering.

☐ **D** They restrict the degradation of chlorophyll. **(1 mark)**

4 Which of the following statements about apical dominance are true?

Statement 1: High levels of auxin directly prevent lateral bud growth.

Statement 2: Cytokinins promote lateral bud growth.

Statement 3: Removing the apical shoot causes the plant to rapidly lose water.

Statement 4: Removing the apical shoot causes the plant to 'bush out'.

☐ **A** only 1 and 2

☐ **C** only 2 and 4

☐ **B** only 2 and 3

☐ **D** only 1, 2 and 4 **(1 mark)**

5 The gibberellin responsible for stem elongation is synthesised by an enzyme coded for by the dominant Le allele. Based on this information, give the genotype for a dwarf pea plant lacking this enzyme.

....

.... **(1 mark)**

Plant responses

1 What is the directional response of plants to light called?

.... **(1 mark)**

2 Plant roots grow away from the light. How does this support plant growth?

.... **(2 marks)**

Practical skills

3 *A student was asked to design an experiment to investigate how growing cress seedlings would respond to unilateral light. The student grew a single batch of cress seedlings lit from above. After a week, she moved the light source to the left.

Note that control variables are needed.

Using your knowledge of auxins, explain what results you would expect and the reasons for this result.

Aim to give a detailed explanation using the information given and your own knowledge. Give evidence to support your points. Make sure your answer is clear and logically structured and that your points are linked.

....

....

....

....

....

....

....

....

....

....

....

.... **(6 marks)**

4 Which of the following is a correct statement about geotropism?

☐ **A** Geotropism is a directional response to gravity.

☐ **B** Geotropism is a directional response to the presence of rocks.

☐ **C** Roots are negatively geotropic.

☐ **D** Shoots are positively geotropic. **(1 mark)**

Commercial use of plant hormones

1 Explain how auxins can be used as an effective broad-leaved weedkiller.

....

....

....

....

....

.... **(3 marks)**

2 Bananas produce a hormone which speeds up the ripening process.

(a) Name this hormone.

.... **(1 mark)**

(b) Describe **one** change in the banana you would expect this hormone to cause.

....

.... **(1 mark)**

(c) This hormone also promotes fruit ripening in cherry trees. Suggest how using the hormone might be of economic benefit to farmers.

....

....

....

....

....

.... **(3 marks)**

3 In wet summers some crop plants can grow too tall, so the plants collapse under the weight of the harvest. Suggest what a farmer could do to prevent this from happening.

Consider which hormone would be responsible for this excessive stem growth.

....

....

....

.... **(2 marks)**

Mammalian nervous system

1 What two structures make up the central nervous system (CNS)?

....

.... **(2 marks)**

2 Which one of the following statements about the mammalian nervous system is false?

☐ **A** Grey matter is made up of non-myelinated neurones.

☐ **B** The peripheral nervous system is made up of both sensory neurones and motor neurones.

☐ **C** All neurones form just one synapse with their neighbouring neurones.

☐ **D** The spinal cord contains both myelinated and unmyelinated neurones. **(1 mark)**

3 In terms of its function, the nervous system can be split into somatic and autonomic systems. Outline the differences between the two.

Make sure that you consider which parts of the body are controlled by each.

....

....

....

....

....

....

....

.... **(4 marks)**

4 The autonomic nervous system can be split into the sympathetic and parasympathetic systems. Which system produces each of the following responses?

	Sympathetic	Parasympathetic
decreased heart rate		
dilation of pupils		
increased blood flow to the brain		
increased blood flow to the digestive system		
increased sexual arousal		
decreased ventilation		

(6 marks)

5 Which of the two autonomic systems is said to be up-regulated during the 'fight or flight' stress response?

.... **(1 mark)**

The brain

1 List **three** processes which the medulla oblongata controls.

....

....

.... **(3 marks)**

2 Which part of the brain is responsible for conscious thought and emotions?

.... **(1 mark)**

3 Which one of the following processes is **not** controlled by the cerebellum?

☐ **A** coordination of relaxation and contraction of antagonistic muscles when walking

☐ **B** forming factual memories

☐ **C** judging the position of objects and limbs when playing a sport

☐ **D** maintaining body position and balance **(1 mark)**

4 Suggest why the cerebellum is sometimes referred to as the body's auto-pilot.

....

....

....

.... **(2 marks)**

5 Outline the role of the hypothalamus in homeostasis.

For 'outline' questions you do not need to make a judgement or explain how or why, you only need give the details requested.

....

....

....

....

....

....

....

.... **(4 marks)**

Reflex actions

1 (a) Suggest what is meant by a **reflect action**?

....

....

....

.... **(2 marks)**

(b) The reflex arc refers to the pathway of neurons involved in a reflex action. This can involve just three neurons. List the types of neurons involved in the correct order.

.... **(3 marks)**

2 Which of the following statements about reflex responses is false?

☐ **A** A reflex response always involves the brain.

☐ **B** Reflex responses are protective.

☐ **C** Reflex responses involve the peripheral and central nervous systems.

☐ **D** Some reflex responses can be overridden by the brain. **(1 mark)**

3 Describe the processes involved in the knee-jerk reflex.

You do not need to make a judgement or explain how or why, you only need to describe what is requested. Make sure you talk about neurones and the direction of impulses in your answer.

....

....

....

....

....

.... **(3 marks)**

4 A number of reflexes have a clear survival value for humans. List **three** examples of protective reflex responses not already mentioned on this page.

These reflexes are seen in newborn children.

....

....

.... **(3 marks)**

Coordination of responses

1 How does the secretion of adrenaline prepare the body for muscular activity?

....

....

....

....

....

....

....

.... **(4 marks)**

2 Adrenaline and cyclic AMP are both messenger molecules. Explain how they perform different but related functions.

....

....

....

.... **(2 marks)**

3 Which one of the following statements about adrenaline is false?

☐ **A** The binding of adrenaline to its receptor activates adenyl cyclase.

☐ **B** Adrenaline is a chemical precursor of cyclic AMP.

☐ **C** Adrenaline relaxes smooth muscles in the bronchioles.

☐ **D** Adrenaline secretion leads to an increase in heart rate. **(1 mark)**

4 Explain how the release of adrenaline from the adrenal gland can bring about responses in a range of tissue types.

What do all target tissues have on the surface of their cell membrane?

....

....

....

.... **(2 marks)**

Controlling heart rate

1 Name the region of the heart responsible for maintaining the resting heart rate at 60–80 beats per minute.

.... **(1 mark)**

2 (a) Which **two** types of receptor provide feedback to the cardiovascular centre in the brain?

....

.... **(2 marks)**

(b) Where are they located?

....

....

....

.... **(2 marks)**

Practical skills

3 Which statement best summarises the control of heart rate in response to exercise?

☐ **A** Heart rate is elevated by an increase in the frequency of impulses passing down the vagus nerve.

☐ **B** The sinoatrial node receives impulses from sensory receptors in the muscles and initiates a greater frequency of waves of excitation across the heart.

☐ **C** An increase in the frequency of impulses from the cardiovascular centre down the accelerator nerve results in an increase in heart rate.

☐ **D** Adrenaline acts by itself to raise heart rate. **(1 mark)**

4 Outline the nervous mechanisms controlling heart rate.

Remember that different nerves are used to increase and decrease heart rate.

....

....

....

....

....

....

.... **(4 marks)**

Had a go ☐ Nearly there ☐ Nailed it! ☐

Muscle structure and function

The diagram shows a sarcomere.

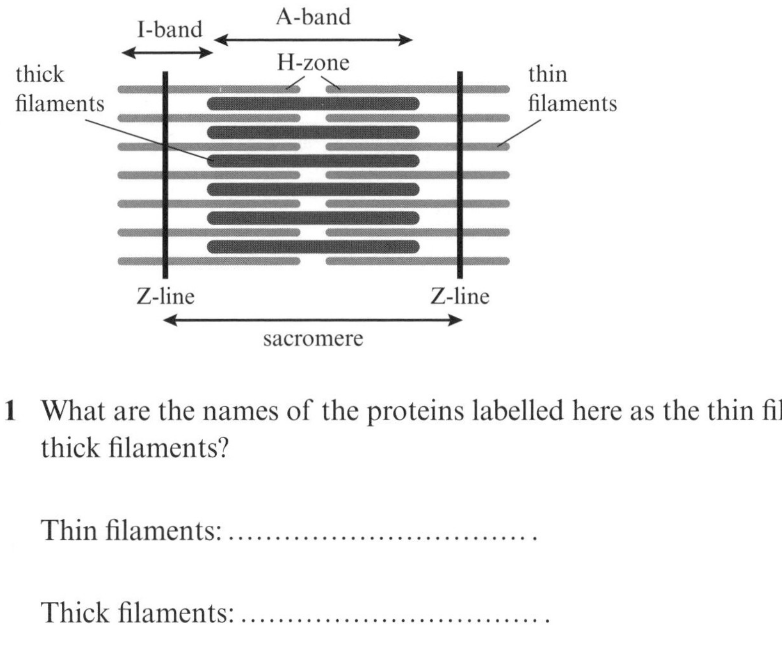

1 What are the names of the proteins labelled here as the thin filaments and thick filaments?

Thin filaments: ....

Thick filaments: .... **(2 marks)**

2 The sarcomere in the diagram is shown during a period of muscle of contraction. Explain what would happen to the following when the muscle relaxes.

(a) The distance between the Z lines

.... **(1 mark)**

(b) The width of the I band

.... **(1 mark)**

(c) The H zone

.... **(1 mark)**

3 Draw an arrow on the diagram pointing to where ATP is hydrolysed during muscle contraction. **(1 mark)**

4 (a) What is a myogenic contraction?

....

.... **(1 mark)**

(b) Which type of muscle can be described as being myogenic?

....

.... **(1 mark)**

5 Where might you find smooth muscle?

....

.... **(1 mark)**

Muscular contraction

1 Outline the processes at the neuromuscular junction that lead to muscle contraction.

> Explain what happens at the synaptic knob, on the muscle fibre membrane and within the muscle fibre.

....

....

....

....

....

....

....

....

....

....

....

.... **(8 marks)**

2 Describe what happens to the level of contraction if a muscle is stimulated to contract repeatedly and at a high frequency.

....

....

....

.... **(2 marks)**

3 Suggest **three** ways in which toxins may interfere with the transmission of a nervous impulse at a neuromuscular junction.

....

....

....

....

....

.... **(3 marks)**

Exam skills

1 The electrical activity of a heart was studied by stimulating the sinoatrial node (SAN). The time taken for electrical activity to be detected in the atrioventricular node (AVN) was recorded. The time taken for electrical activity to be detected in the SAN, at the start of the next cycle, was also recorded.

Area of the heart	Time when electrical activity was detected (s)
AVN	0.15
SAN	0.75

Remember the SAN produces electrical activity intrinsically.

(a) Describe what happens in the atria during the 0.15 seconds before the AVN shows any electrical activity.

....

....

....

....

....

.... **(3 marks)**

Maths skills

(b) Use the information in the table to calculate how often the SAN would show electrical activity in 1 minute. Show your working.

.... **(2 marks)**

2 Which of the following statements is correct?

☐ **A** The sympathetic nervous system is a branch of the somatic system and prepares the body for fight or flight.

☐ **B** The sympathetic nervous system is a branch of the autonomic system and prepares the body for fight or flight.

☐ **C** The sympathetic nervous system is a branch of the somatic system and prepares the body for rest and digest.

☐ **D** The sympathetic nervous system is a branch of the autonomic system and prepares the body for rest and digest. **(1 mark)**

3 Which of the following sequences best represents the length of time over which each of these hormones usually acts.

> means 'greater than'

☐ **A** insulin > adrenaline > testosterone

☐ **B** adrenaline > testosterone > insulin

☐ **C** testosterone > insulin > adrenaline

☐ **D** All act over the same length of time. **(1 mark)**

Photosynthesis and respiration

1 Give the full chemical symbol equation for photosynthesis.

....

.... **(2 marks)**

2 In which organelle do the reactions of photosynthesis take place?

.... **(1 mark)**

3 In terms of respiration and photosynthesis, explain the difference between autotrophs and heterotrophs.

....

....

....

.... **(2 marks)**

4 Give two similarities between photosynthetic bacteria and chloroplasts.

Both are able to synthesise their own proteins.

....

....

....

.... **(2 marks)**

5 The endosymbiont theory suggests that photosynthetic bacteria were absorbed through endocytosis by early eukaryotic cells. What advantages would there be for the eukaryotic cells in this behaviour?

....

....

....

.... **(2 marks)**

6 The products of photosynthesis are a solute and a gas. Outline the processes which allow each to leave the organelle responsible for photosynthesis.

....

....

....

.... **(2 marks)**

Photosystems, pigments and thin layer chromatography

1 The diagram shows a chloroplast.

(a) Identify the features:

W ....

X ....

Y ....

Z ....

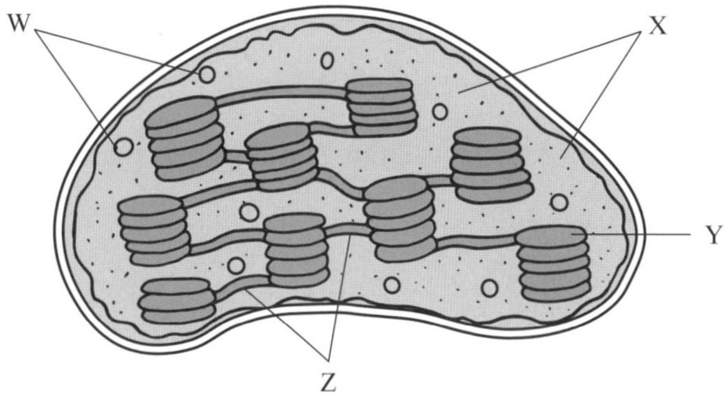

(4 marks)

(b) In which structure would you expect find the greatest concentration of chlorophyll?

.... **(1 mark)**

(c) How does the structure of Y support photosynthesis?

....

.... **(1 mark)**

(d) Suggest how X differs in content to the cytoplasm of the cell.

....

.... **(1 mark)**

2 Define the term **photosystem**.

....

....

....

....

....

.... **(3 marks)**

3 Indicate which of the following are associated with chlorophyll a, chlorophyll b or carotenoids.

	Chlorophyll a	Chlorophyll b	Carotenoids
found in two forms: P700 and P680			
reflects orange and yellow light			
contains a porphyrin ring			
appears blue-green			
accessory pigments			

(5 marks)

Light-dependent stage

1 The first part of the light-dependent stage of photosynthesis can be summarised in the equation $2H_2O \rightarrow 4H^+ + 4e^- + O_2$

(a) What is the name given to this process?

.... **(1 mark)**

(b) Suggest what might happen to the oxygen produced during this process.

....

....

....

.... **(2 marks)**

(c) Where exactly in the chloroplast does this process take place?

....

.... **(1 mark)**

2 Photophosphorylation is the production of ATP using light energy.

Which sequence presents the correct order of events?

1 A photon of light strikes photosystem II.

2 Electrons from photosystem I combine with protons to reduce NADP to reduced NADP.

3 Electrons lost from photosystem I are replaced by those from photosystem II.

4 Two electrons from photosystem II are lost.

5 Electrons move along the electron transport chain between photosystems.

☐ **A** 1, 2, 3, 4, 5

☐ **B** 1, 3, 2, 4, 5

☐ **C** 1, 4, 5, 3, 2

☐ **D** 1, 4, 2, 5, 3 **(1 mark)**

3 Outline the differences between cyclic and non-cyclic photophosphorylation.

....

....

....

....

....

....

....

....

....

.... **(5 marks)**

Light-independent stage

1 What is the name given to the cycle of reactions which make up the light-independent stage (LIS) of photosynthesis?

.... **(1 mark)**

2 The following sketch shows a simplified version of the LIS of photosynthesis.

(a) Identify substances X and Y.

X ....

Y .... **(2 marks)**

(b) Which of the substances shown in the sketch would normally be used in the synthesis of glucose?

.... **(1 mark)**

(c) Name the enzyme that combines ribulose bisphosphate with substance X to produce substance Y.

.... **(1 mark)**

3 Which one of the following statements about the LIS of photosynthesis is false?

☐ **A** Fatty acids can be made from one of the LIS intermediates.

☐ **B** The LIS takes place in the stroma.

☐ **C** The LIS takes place mainly in the thylakoids of the chloroplast.

☐ **D** The products of the LDS of photosynthesis are directly used in the LIS. **(1 mark)**

4 The enzyme referred to in question 2(c) has an optimum pH of around 8. This enzyme acts in the stroma of the chloroplast. Why does the stroma have a pH of 8?

....

....

....

.... **(2 marks)**

5 Name **two** products of the LDS that are used in the LIS.

....

....

....

.... **(2 marks)**

Factors affecting photosynthesis

1 Define the term **limiting factor**.

....

.... **(1 mark)**

Maths skills

2 The sketch shows a curve which could be used to explain the effects of an increase in light intensity on the rate of photosynthesis.

Rate of photosynthesis

Y

X

Light intensity

With reference to limiting factors, describe and explain the shape of the graph at points X and Y.

X: ....

....

....

....

Y: ....

....

....

.... **(4 marks)**

Guided

3 *Explain how greenhouses can be used to maximise the yield of certain crop plants such as tomatoes.

> A longer-answer question requires a good structure. Consider each of the limiting factors and how they could be controlled – what systems could be used?

Carbon dioxide, temperature and light intensity are all factors which affect the rate of photosynthesis. Carbon dioxide levels can be increased by ....

....

....

....

....

....

....

....

....

.... **(6 marks)**

The need for cellular respiration

1 Explain what is meant by the term **respiration**.

....

....

....

.... **(2 marks)**

2 Explain the differences between anabolic and catabolic reactions.

This is an 'Explain' question, so it is not enough to state your point - you must also give reasons or say why you have come to that conclusion.

....

....

....

.... **(2 marks)**

3 Besides active transport and cell division, give **three** more roles of ATP within cells.

....

....

....

....

....

.... **(3 marks)**

4 Which of the following is **not** associated with active transport?

☐ **A** movement against a concentration gradient

☐ **B** movement with a concentration gradient

☐ **C** the use of ATP

☐ **D** the use of transport proteins **(1 mark)**

5 Draw a diagram in the space below to show the structure of ATP.

ATP is adenosine triphosphate.

(3 marks)

Glycolysis

1 Where does glycolysis take place in cells?

....

.... **(1 mark)**

2 This is a simplified version of the glycolytic pathway.

glucose
X
2 × triose sugar
Y
2 × pyruvate

(a) At which stage is ATP used?

....

(1 mark)

(b) At which stage is ATP produced?

....

(1 mark)

(c) How many molecules of ATP are made during the stage you have indicated in part (b)?

....

.... **(1 mark)**

3 Glycolysis produces ATP and pyruvate. Name another molecule produced by glycolysis that is important in respiration.

.... **(1 mark)**

4 Which one of these statements about glycolysis is **not** true?

☐ **A** Glycolysis can proceed in the absence of oxygen.

☐ **B** Glycolysis is an ancient biochemical pathway shared by all prokaryotic and eukaryotic cells.

☐ **C** Glycolysis is the same process in mammals and in yeast.

☐ **D** Glycolysis uses more ATP than it makes. **(1 mark)**

5 A redox reaction is a reaction where one molecule becomes reduced and one becomes oxidised. Explain the significance of redox reactions in glycolysis.

....

....

....

.... **(2 marks)**

Structure of the mitochondrion

1 Here is a diagram of a mitochondrion.

Identify parts X, Y and Z.

X ....

Y ....

Z ....

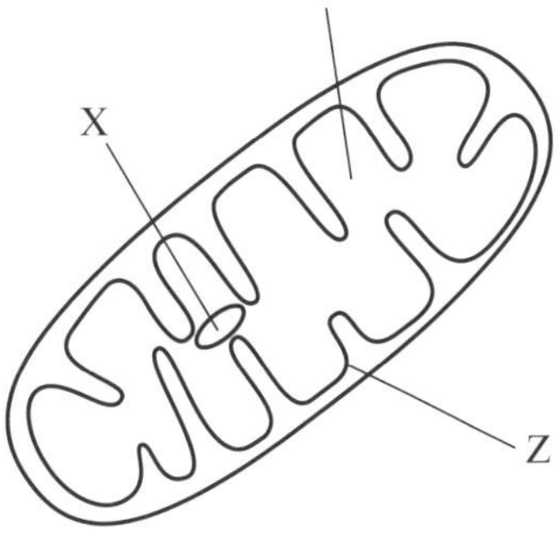

(3 marks)

2 The link reaction requires pyruvate.
How does pyruvate enter a mitochondrion?

....

....

....

.... **(2 marks)**

3 Which statement about mitochondria is true?

☐ **A** They have a double membrane.

☐ **B** They are always spherical in shape.

☐ **C** Their outer and inner membrane are both permeable to protons.

☐ **D** They contain cytoplasm. **(1 mark)**

4 Explain how the inner membrane of the mitochondrion is adapted to its function.

The inner membrane has a number of important proteins embedded in it.

....

....

....

....

....

....

....

.... **(4 marks)**

Link reaction and the Krebs cycle

1 Where in the cell does the link reaction and the Krebs cycle take place?

.... **(1 mark)**

2 Which statement best describes the link reaction?

☐ **A** a condensation reaction involving pyruvate and carbon dioxide

☐ **B** a reduction reaction where reduced NAD contributes hydrogen atoms to pyruvate making acetyl CoA

☐ **C** the combining of two pyruvate molecules to make acetyl CoA

☐ **D** the conversion of pyruvate to acetyl CoA involving a decarboxylation step followed by dehydrogenation involving the coenzyme NAD **(1 mark)**

3 The diagram shows a simplified version of the Krebs cycle.

acetyl CoA

X

Y

α-ketoglutarate
(5 carbons)

(a) Identify compounds X and Y.

X................................

Y................................ **(2 marks)**

(b) How many carbon atoms are there in compound Y?

.... **(1 mark)**

(c) Indicate on the diagram where you expect carbon dioxide to be produced. **(2 marks)**

4 Complete the table below to show how many molecules of reduced NAD, reduced FAD, ATP and carbon dioxide are made during the link reaction and the Krebs cycle for every one glucose molecule respired.

	Number of made in link reaction	**Number of made in Krebs cycle**
reduced NAD		
reduced FAD		
ATP		
CO_2		

(4 marks)

Oxidative phosphorylation

1 Where precisely in the cell does oxidative phosphorylation take place?

....

.... **(1 mark)**

2 Substrate-level phosphorylation involves the production of ATP whilst the substrate remains involved in the reaction process.

(a) During which stages of respiration is ATP produced in this way?

....

.... **(2 marks)**

(b) How many molecules of ATP are produced via substrate-level phosphorylation from one glucose molecule?

....

.... **(2 marks)**

3 Which one of the following statements about oxidative phosphorylation is false?

☐ **A** Just two of the complexes in the electron transport chain (ETC) pump protons.

☐ **B** Protons flow through ATP synthase and drive the production of ATP.

☐ **C** The electrons moving along the ETC originate from molecules of reduced NAD.

☐ **D** The pumping of protons decreases the pH in the matrix of the mitochondria. **(1 mark)**

4 Cytochrome oxidase is an enzyme found at the end of the electron transport chain. Describe the role of this enzyme.

....

....

....

....

....

.... **(3 marks)**

Anaerobic respiration in eukaryotes

1 In the absence of oxygen, eukaryotic organisms respire anaerobically. Which one of the four stages of respiration can be described as anaerobic?

☐ **A** glycolysis
☐ **C** link reaction
☐ **B** Krebs cycle
☐ **D** oxidative phosphorylation **(1 mark)**

2 Reduced NAD is reoxidised to NAD in oxidative phosphorylation. In the absence of oxygen this does not happen. In mammals, an alternative hydrogen acceptor is used to reoxidise reduced NAD.

(a) What is this hydrogen acceptor called?

.... **(1 mark)**

(b) What is the name of the enzyme which catalyses this reaction?

.... **(1 mark)**

3 When people exercise vigorously, lactate can build up in muscle tissue.

(a) Explain what happens to this lactate in mammals.

....

....

....

.... **(2 marks)**

(b) What would occur in the muscle tissue if the lactate remained?

How might lactate (lactic acid) affect the pH of the tissue fluid in muscles?

....

.... **(1 mark)**

Maths skills

4 Fill in the gaps in the diagram to complete the equation for anaerobic respiration in yeast cells.

Pyruvate → (a) .... → (b) ....
(RedNAD → NAD)
Pyruvate → (c) ....

(3 marks)

5 Indicate whether each statement applies to anaerobic respiration in yeast, mammals or both.

	Yeast	Mammals	Both
Ethanal is the hydrogen acceptor.			
Carbon dioxide is produced.			
Reduced NAD is reoxidised.			
Lactate is made.			
Enzymes are required.			

(5 marks)

Energy values of different respiratory substrates

1 Define the term **respiratory substrate**.

....

.... **(1 mark)**

Maths skills

2 (a) Different respiratory substrates have different energy values, related to the number of molecules of ATP they yield per unit mass. Match the following substrates with their mean energy value (given in kJ g^{-1}).

Substrate	Mean energy value (given in kJ g^{-1})
carbohydrate	
lipid	
protein	

Mean energy values (given in kJ g^{-1}): 15.8, 17.0, 39.4 **(3 marks)**

(b) What determines the amount of ATP that each of these respiratory substrates can produce?

....

....

.... **(2 marks)**

3 Which one of the following statements is true?

☐ **A** Glucose provides the richest source of energy of all biological molecules.

☐ **B** Many amino acids can be metabolised, and the products fed into the Krebs cycle can be respired.

☐ **C** Only the fatty acid portion of lipids can be respired to release energy.

☐ **D** When the body is undergoing starvation, protein is converted into fat before it can be respired. **(1 mark)**

Maths skills

4 The respiratory quotient (RQ) is calculated using the following formula:

$$RQ = \frac{\text{carbon dioxide produced}}{\text{oxygen used}}$$

In terms of the volumes of oxygen used and carbon dioxide produced, what would an RQ value of 1 mean?

....

....

....

.... **(1 mark)**

Factors affecting respiration

1 Explain how an increase in temperature might affect the rate of aerobic respiration inside liver cells.

Try to give specific examples of how an increase in temperature would impact on a particular process, e.g. diffusion of H^+ ions through ATP synthase.

....

....

....

....

....

....

....

.... **(4 marks)**

2 Which one of the following statements is true?

☐ **A** All plant cells respire but not all plant cells can carry out photosynthesis.

☐ **B** All stages of respiration take place in the mitochondrion.

☐ **C** Respiration relies on the action of enzymes to proceed.

☐ **D** The rate of respiration is affected by a change in light intensity. **(1 mark)**

Practical skills

3 A respirometer is a piece of laboratory equipment which can be used to help measure the rate of respiration.

(a) What does this equipment actually measure?

....

.... **(1 mark)**

Maths skills

(b) What other piece of equipment would you need in order to calculate a rate?

.... **(1 mark)**

(c) What units might you use when presenting these data?

.... **(1 mark)**

Practical skills

4 Sodium hydroxide is used in respirometers to absorb the carbon dioxide produced during respiration. How does this help determine the rate of respiration?

....

.... **(1 mark)**

Exam skills

1 The diagram below shows some of the steps involved in photosynthesis.

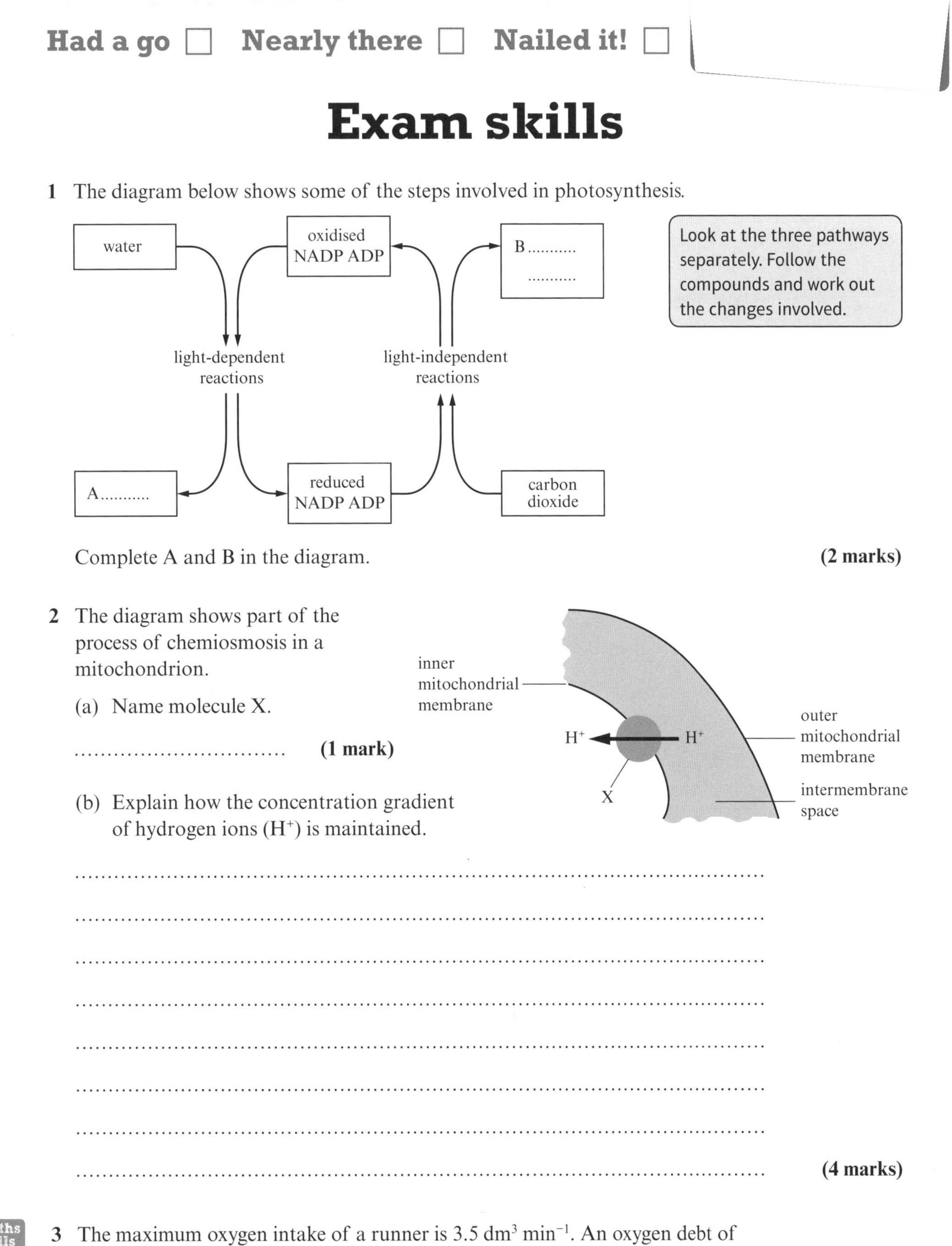

Look at the three pathways separately. Follow the compounds and work out the changes involved.

Complete A and B in the diagram. **(2 marks)**

2 The diagram shows part of the process of chemiosmosis in a mitochondrion.

(a) Name molecule X.

.... **(1 mark)**

(b) Explain how the concentration gradient of hydrogen ions (H^+) is maintained.

....

....

....

....

....

....

....

.... **(4 marks)**

Maths skills

3 The maximum oxygen intake of a runner is 3.5 $dm^3\ min^{-1}$. An oxygen debt of 14.5 dm^3 can be incurred. The runner needs oxygen at a rate of 0.3 $dm^3\ s^{-1}$ to run at 5 $m\ s^{-1}$. Calculate the maximum distance the runner can cover before being overcome by exhaustion.

Start by calculating the runner's oxygen requirement in dm^3 per minute and compare that to their maximum oxygen intake per minute. Make sure oxygen intake values are all in the same units.

.... **(4 marks)**

Gene mutation

1 What is meant by the term **gene mutation**?

.... **(1 mark)**

2 What type of gene mutation alters the gene sequence by adding a single nucleotide?

.... **(1 mark)**

Guided

3 Part of the gene sequence for an enzyme is shown below. The second sequence shows the result of a mutation.

CGGAATGGGTAG → CGGAATTGGGTAG

Ala Leu Pro Ile → Ala Leu Thr His

(a) Describe what has happened to the reading frame of the gene sequence and explain the consequences of this.

Compare the two sequences, both base by base and in blocks of three.

The reading frame of the gene sequence has been shifted by one nucleotide, causing a frameshift. ....

....

....

....

....

....

....

....

.... **(5 marks)**

(b) Suggest what might happen to the functionality of the enzyme in part (a).

Think about how a change in the amino acid sequence could affect the active site of the enzyme.

....

....

....

....

....

.... **(3 marks)**

Gene control

1 What is the difference between a structural gene and a regulatory gene?

....

....

....

.... **(2 marks)**

2 The diagram below shows the *lac* operon.

regulatory gene		promoter	operator	structural gene Z	structural gene Y

The *E. coli lac* operon and its regulatory gene.

(a) Describe how genes Z and Y are controlled in the absence of lactose.

> The regulatory gene produces a repressor protein.

....

....

....

....

....

.... **(3 marks)**

(b) Describe how genes Z and Y are controlled in the presence of lactose.

....

....

....

.... **(2 marks)**

(c) Comment on why the operon genes are controlled in this way.

> You need to present an informed opinion or comment on relevant points of interest.

....

....

....

....

....

.... **(3 marks)**

Homeobox genes

1 (a) What is a **homeobox gene**?

....

....

....

.... **(2 marks)**

(b) Describe how the homeobox genes are arranged.

....

....

....

.... **(2 marks)**

(c) Explain how a homeobox gene carries out its function.

This is an 'explain question' so set out reasons using your biological knowledge of how transcription factors work.

....

....

....

....

....

....

....

.... **(4 marks)**

2 Homeobox genes are highly conserved across species.

(a) What does **highly conserved across species** mean?

....

.... **(1 mark)**

(b) Suggest why these genes have not been changed by mutation.

....

....

....

.... **(2 marks)**

Mitosis and apoptosis

1 (a) Describe how mitosis is used in the body.

....

....

....

.... **(2 marks)**

Guided

(b) Place the stages of mitosis in the correct order.

Do you have a mnemonic for the stages of mitosis?

prophase

....

....

.... **(3 marks)**

2 (a) Outline the changes that happen to cells during apoptosis.

Include the term 'bleb' in your answer and restrict the outline to essential detail only.

....

....

....

....

....

....

....

.... **(4 marks)**

(b) What happens to the remains of cells that have undergone apoptosis?

....

.... **(1 mark)**

3 Skin cells were observed in the process of mitosis. Outline what you would expect to see in a cell that was going through metaphase.

....

....

.... **(2 marks)**

Variation

1 (a) What is meant by **continuous variation**?

....

.... **(1 mark)**

(b) What factors influence continuous variation?

1 ....

2 .... **(2 marks)**

(c) Humans have four different possible blood groups. What type of variation is this?

.... **(1 mark)**

2 (a) Describe how genetic variation could occur in a species.

Include the terms 'independent assortment' and 'crossing over' in your answer.

For 'describe questions', you do not need to make a judgement or explain how or why, you only need to describe what is requested.

....

....

....

....

....

....

....

....

....

.... **(5 marks)**

(b) Suggest how any mutations could be spread throughout a bacterial population.

....

....

....

.... **(2 marks)**

Inheritance

1 Sickle cell disease is caused by a recessive allele, Hb^S. (Hb^N is used for the normal allele.)

(a) What word is used to describe a person with one copy of the recessive allele?

.... **(1 mark)**

Guided

(b) Two parents each have a single copy of the recessive allele. What possible gametes could they produce?

Mother: Hb^S Hb^N

Father: .... **(1 mark)**

(c) What are the possible genotypes of their children?

....

.... **(2 marks)**

Maths skills

(d) Calculate the probability that they will have a child with sickle cell disease?

Probability is worked out as a percentage. For example, a 50/50 probability could be written as 0.5 or 50%. Show your working.

.... **(1 mark)**

2 A species of plant can have red or white petals that are smooth or frilled. Red (R) is dominant to white (r) and smooth (F) is dominant to frilled (f). These genes are not linked.

Two plants with smooth red petals were bred together and produced offspring.

(a) Complete the table below to show the genotypes of the offspring.

	RF	Rf	rF	rf
RF				
Rf				
rF				
rf				

(4 marks)

(b) What are the phenotypes of the offspring?

The phenotype is how the organism looks. For example, the genotype Rrff gives a phenotype of red, frilled petals.

....

.... **(2 marks)**

Maths skills

(c) What is the ratio of the phenotypes?

.... **(1 mark)**

Linkage and epistasis

1 Colour blindness is a recessive trait that is carried on the X chromosome, X^b.

(a) What type of inheritance is this?

.... **(1 mark)**

Guided

(b) A woman with normal colour vision but who carries the allele for colour blindness has a child with a man who is colour blind. What are the possible genotypes of their child?

First work out the genotypes of the two parents.

X^BX^b .... **(2 marks)**

Maths skills

(c) What is the probability that their child is colour blind?

.... **(1 mark)**

2 Coat colour in Labrador dogs is controlled by epistatic genes. Coat colour black (B) is dominant to coat colour chocolate (b), but only in the presence of the dominant allele for the fur colour gene, E. In the absence of a dominant allele for fur colour, the coat colour is yellow.

(a) What possible genotypes could a black Labrador have?

.... **(2 marks)**

(b) A chocolate Labrador that is heterozygous for the fur colour gene breeds with a yellow Labrador that is homozygous recessive for both genes. Show the genetic cross.

The genotype of this chocolate Labrador is bbEe. What are the possible genotypes for a yellow Labrador?

(4 marks)

Maths skills

(c) What is the probability of the cross from part (b) producing yellow offspring?

.... **(1 mark)**

Using the chi-squared test

A researcher has counted the number of plants in a shaded and an unshaded area of forest. The results are shown below.

Number of plants in shaded area	Number of plants in unshaded area
68	96

1 Suggest a null hypothesis.

....

.... **(1 mark)**

Guided
Maths skills

2 The formula for χ^2 is $\chi^2 = \Sigma \frac{(\text{observed} - \text{expected})^2}{\text{expected}}$

Calculate the χ^2 value.

The χ^2 value must be greater than the critical value at $p = 0.05$ for the data to be considered significant.

	O	E	(O − E)	$(O - E)^2$	$\frac{(O - E)^2}{E}$
Shaded area	68	82	....	196	....
Unshaded area	96	82	14	....	....
				$\sum=$	....

(3 marks)

Maths skills

3 Use the χ^2 critical values table below to explain if the numbers of plants in the two areas are significantly different.

The accepted critical value for significance is $p = 0.05$, although you should also check your value against the critical values lower than $p = 0.05$ (e.g. $p = 0.025$).

df	0.25	0.10	0.05	0.025	0.01
1	1.323	2.706	3.841	5.024	6.635
2	2.773	4.605	5.991	7.378	9.210
3	4.108	6.251	7.815	9.348	11.345

....

....

....

....

....

.... **(3 marks)**

4 Suggest why there are more plants in the unshaded area.

....

.... **(1 mark)**

The evolution of a species

1 The purple-throated mountain gem and the white-throated mountain gem are two closely related species of hummingbird from the mountains of North America. Typically, white-throated mountain gems are found at higher altitudes than purple-throated mountain gems.

(a) Name the process that has given rise to these two species from their common ancestor.

.... **(1 mark)**

(b) Outline a mechanism for the formation of these two species.

Suggest a type of isolation that could separate two breeding populations.

....

....

....

....

....

....

....

.... **(4 marks)**

(c) What is this type of speciation called?

.... **(1 mark)**

2 A species of snail has three phenotypes, black, grey and white. A particular population of the snails lives in a cold, grey, rocky environment and is predated by small birds.

(a) What is the selective pressure on the snails?

....

.... **(1 mark)**

(b) There are more grey snails than black snails or white snails in the population. Suggest why.

....

....

....

.... **(2 marks)**

(c) What is this type of selection called?

.... **(1 mark)**

The Hardy–Weinberg principle

Maths skills

1 A species of ladybird has two different phenotypes, red with black spots or orange with black spots, controlled by one gene. The allele for red with black spots is dominant (R) and the allele for orange with black spots is recessive (r).
In a particular population, 22 per cent of the ladybirds are orange with black spots.

> The equations you need to use are:
> $p + q = 1$ $p^2 + 2pq + q^2 = 1$

Guided

(a) What percentage of the population is homozygous for the dominant allele?

$q^2 = 0.22$, therefore $q = 0.469$

................................% **(2 marks)**

(b) What percentage of the population is heterozygous?

................................% **(1 mark)**

(c) What is the frequency of the dominant allele in the population?

.... **(1 mark)**

Maths skills

2 In Australia, 1 in 10 000 people suffer from phenylketonuria, a disease caused by a recessive allele.

(a) What percentage of the population are carriers for phenylketonuria?

> The percentage that corresponds to 1 in 10 000 will be the percentage of people who are homozygous (q^2).

....

....

....

....

....

.... **(3 marks)**

(b) If the population of Australia is 24 000 000, how many people suffer from phenylketonuria?

....

.... **(1 mark)**

(c) What is the frequency of the recessive allele in Australia?

....

.... **(1 mark)**

Artificial selection

1 Humans have been selectively breeding cattle for milk for thousands of years.

(a) State **three** characteristics selected for in dairy cows.

1 ....

2 ....

3 .... **(3 marks)**

(b) Suggest a mechanism for the selective breeding of dairy cows.

Here you need to draw on your biological knowledge to answer this question.

....

....

....

.... **(2 marks)**

Guided

(c) Modern reproductive technology has been used to assist artificial selection in recent years. Name **three** modern procedures that are used to breed dairy cows.

1 in-vitro ....

2 ....

3 .... **(3 marks)**

2 Conservation projects worldwide protect species by placing animals into nature reserves and plants into botanical gardens.

(a) Explain why it is important for artificial selection that we preserve species.

....

.... **(1 mark)**

(b) Discuss the ethical considerations that should be made when using animals for selective breeding.

Aim to give a detailed account that addresses both sides of the argument.

....

....

....

.... **(2 marks)**

Exam skills

1 The diagram below shows the inheritance of haemophilia in a family.

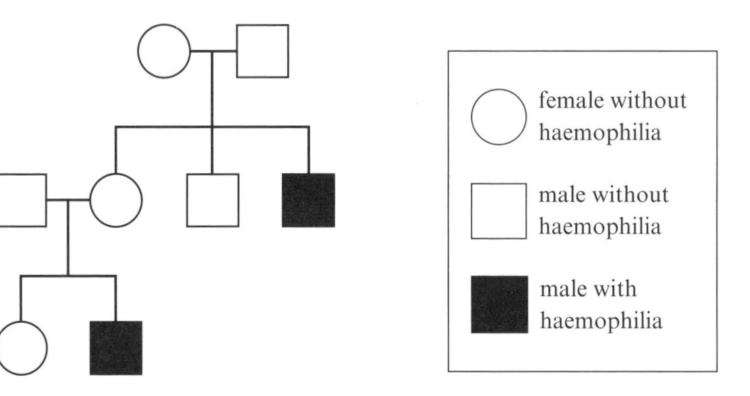

(a) Describe how the male sufferers inherited haemophilia.

Haemophilia is a recessive trait.

....

....

....

....

....

.... **(3 marks)**

Maths skills

(b) What is the probability that the youngest female is a carrier?

................................% **(2 marks)**

(c) *Using evidence from the diagram above to justify your answer, explain why males are more likely to suffer from haemophilia than females.

....

....

....

.... **(6 marks)**

DNA sequencing

1 (a) What is **gene sequencing**?

....

.... **(1 mark)**

(b) Describe how enzymes are involved in gene sequencing.

....

....

....

.... **(2 marks)**

(c) A DNA template is sequenced using A, C and G nucleotides and a T dideoxynucleotide. Use the DNA template below to work out the complementary DNA fragment.

> Use complementary base pairing.

TTGCCCGAGGGG

.... **(1 mark)**

(d) Explain how a computer reads the DNA sequence from an electrophoresis gel.

....

....

....

.... **(2 marks)**

2 (a) Name **two** uses of gene sequencing in biology.

1 ....

2 .... **(2 marks)**

Guided (b) New gene sequencing techniques are currently used to speed up the process. Describe one new technique.

DNA templates are attached to microbeads and many copies are

made using ....

....

.... **(2 marks)**

Polymerase chain reaction

1 (a) What is the polymerase chain reaction (PCR)?

....

.... **(1 mark)**

(b) Describe the process of PCR.

There are three main stages, each with a particular temperature.

....

....

....

....

....

....

....

....

....

.... **(5 marks)**

2 (a) Explain how PCR is used in genetic fingerprinting.

....

....

....

....

....

.... **(3 marks)**

(b) What conditions make it difficult for PCR to be used?

....

....

....

.... **(2 marks)**

Gel electrophoresis

Practical skills

1 (a) Describe the process of gel electrophoresis.

....

....

....

....

....

....

....

.... **(4 marks)**

(b) Name **two** processes that use gel electrophoresis.

1 ....

2 .... **(2 marks)**

Guided

2 Each person's genetic fingerprint is different due to the presence of VNTRs.

(a) What is a VNTR?

VNTR stands for variable .... **(1 mark)**

....

(b) Explain how VNTRs result in fragments of DNA of different sizes.

....

....

....

....

....

.... **(3 marks)**

(c) Identify some reasons why people's genetic fingerprints are similar to the genetic fingerprints of their parents.

Select relevant points from your knowledge of genetic fingerprints.

....

....

....

.... **(2 marks)**

Genetic engineering

1 (a) What is **genetic engineering**?

....

.... **(1 mark)**

Guided

(b) Name **two** examples of uses of genetic engineering.

1 making human insulin from ....

2 .... **(2 marks)**

(c) Describe **two** ways to isolate a gene of interest from an organism's DNA.

Discuss two different methods.

....

....

....

....

....

....

....

.... **(4 marks)**

2 (a) What is **recombinant DNA**?

.... **(1 mark)**

(b) Describe how enzymes are used in recombinant plasmids.

....

....

....

.... **(2 marks)**

(c) Explain why sticky ends are important in combining two sections of DNA.

....

....

....

.... **(2 marks)**

The ethics of genetic manipulation

1 (a) Give **two** examples of how we use genetically modified plants.

1 ....

2 .... **(2 marks)**

(b) What is **pharming**?

....

.... **(1 mark)**

(c) Give one ethical reason **for** and one ethical reason **against** using genetically modified animals.

Think about who might suffer if you did not use genetically modified animals and microorganisms.

For ....

....

Against ....

.... **(2 marks)**

2 Genetically modified rice contains beta carotene. Why has it been developed?

....

....

....

.... **(2 marks)**

3 What are the ethical problems with having patented seeds?

....

....

....

.... **(2 marks)**

4 Discuss the possible implications of growing genetically modified plants in an open field near to other plant species.

....

....

....

.... **(2 marks)**

Gene therapy

1 (a) What is **germ line gene therapy**?

....

....

....

.... **(2 marks)**

(b) Suggest an ethical reason for not using this type of gene therapy.

....

.... **(1 mark)**

2 SCID is a genetic disease that has been cured using somatic cell gene therapy, using a virus as a vector to place a functional allele into bone marrow cells.

(a) What is **SCID**?

....

....

....

.... **(2 marks)**

(b) Give a method, other than using a virus, of putting a functional allele into the correct body cells. In your answer give the vector and method of delivery.

....

....

....

.... **(2 marks)**

Guided

(c) Explain some of the problems associated with somatic cell gene therapy.

It is difficult to get the functional allele into the right cells.

....

....

....

....

.... **(2 marks)**

Exam skills

Practical skills

1 A group of researchers are investigating two species of lemur in Madagascar and are taking blood samples in order to extract DNA.

(a) Only small amounts of DNA can be extracted from the blood. Describe how the DNA sample could be amplified.

Discuss the process of the polymerase chain reaction.

(3 marks)

(b) The researchers want to find out how closely related the two species of lemur are. What technique should they use?

(1 mark)

(c) (i) One of the female lemurs in one family group has a baby. Explain how the researchers could use genetic fingerprinting to show which one of the male lemurs is the father.

(4 marks)

(ii) The diagram shows a genetic fingerprint of the mother, baby and possible fathers (A, B or C). Explain which column shows the father of the baby.

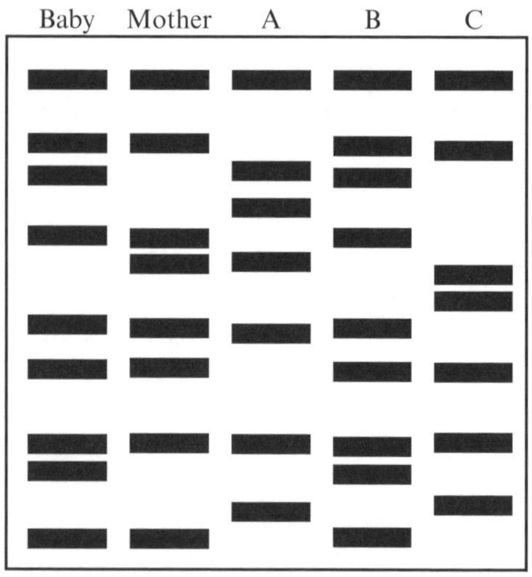

(2 marks)

Baby Mother A B C

Natural clones

1 Clones occur naturally in nature. Give **two** examples of natural cloning in plants.

....

....

....

.... **(2 marks)**

2 What type of reproduction is cloning?

.... **(1 mark)**

3 Describe how plant cloning is used in horticulture to produce more plants.

....

....

....

....

....

.... **(3 marks)**

4 *Give a detailed evaluation of the advantages and disadvantages of cloning plants using your own knowledge.

> Note the number of marks: aim to give **three** advantages and **three** disadvantages. Give evidence to support your points. Make sure your answer is clear and logically structured and that your points are linked.

....

....

....

....

....

....

....

....

....

....

....

.... **(6 marks)**

Artificial clones

1 A horticulturist wants to produce many clones from one parent plant using micropropagation. Describe the process of micropropagation.

....

....

....

....

....

.... **(3 marks)**

2 Give **one** disadvantage of micropropagation.

....

.... **(1 mark)**

3 Recently, cloning of farm animals has been used to produce many animals with the desired characteristics.

(a) What is **artificial embryo twinning**?

....

.... **(1 mark)**

Guided

(b) Describe the process of artificial embryo twinning.

A developing embryo is divided into separate cells. ....

....

....

....

.... **(2 marks)**

4 (a) Give **one** advantage of animal cloning.

....

.... **(1 mark)**

(b) Give **one** disadvantage of animal cloning.

....

.... **(1 mark)**

Microorganisms in biotechnology

1 What is **biotechnology**?

....

.... **(1 mark)**

2 Give **two** examples of the uses of microorganisms in biotechnology.

1 ....

2 .... **(2 marks)**

3 *Some students investigated the optimum conditions needed to grow bacteria in a flask. The number of bacteria was shown by the opacity of the flask. The opacity of the flask was estimated using standards, 0%, 2%, 5%, 10% and 20%. The results are shown below:

Make sure you structure your answer logically, showing how the points you make are related or follow on from each other where appropriate. You should also support your points with relevant biological facts and/or evidence.

Temperature / °C	pH	Opacity / %
20	5	2
40	5	5
60	5	2
20	7	5
40	7	20
60	7	5

The students' tutor had the following concerns about their method of data collection and their results:

1 The method was not a valid test of what was being investigated.

2 The results may not be accurate.

3 The results were not reliable.

Explain why these concerns are justified and suggest improvements to the investigation.

....

....

....

....

....

.... **(3 marks)**

Aseptic techniques

1 (a) What does **aseptic** mean?

.... **(1 mark)**

(b) List **two** aseptic techniques.

1 ....

2 .... **(2 marks)**

2 Fermentation requires aseptic techniques.

(a) Compare the two different types of fermentation.

Name the two types of fermentation and then describe them.

....

....

....

....

....

....

....

.... **(4 marks)**

(b) Explain why aseptic techniques are so important inside a fermenter.

....

....

....

.... **(3 marks)**

3 Penicillin is made in a fermenter.

(a) What type of fermentation produces penicillin?

....

.... **(1 mark)**

(b) Explain why this type of fermentation is needed to make penicillin.

....

....

....

.... **(2 marks)**

Growth curves of microorganisms

The graph below shows the growth curve of microorganisms in a batch culture.

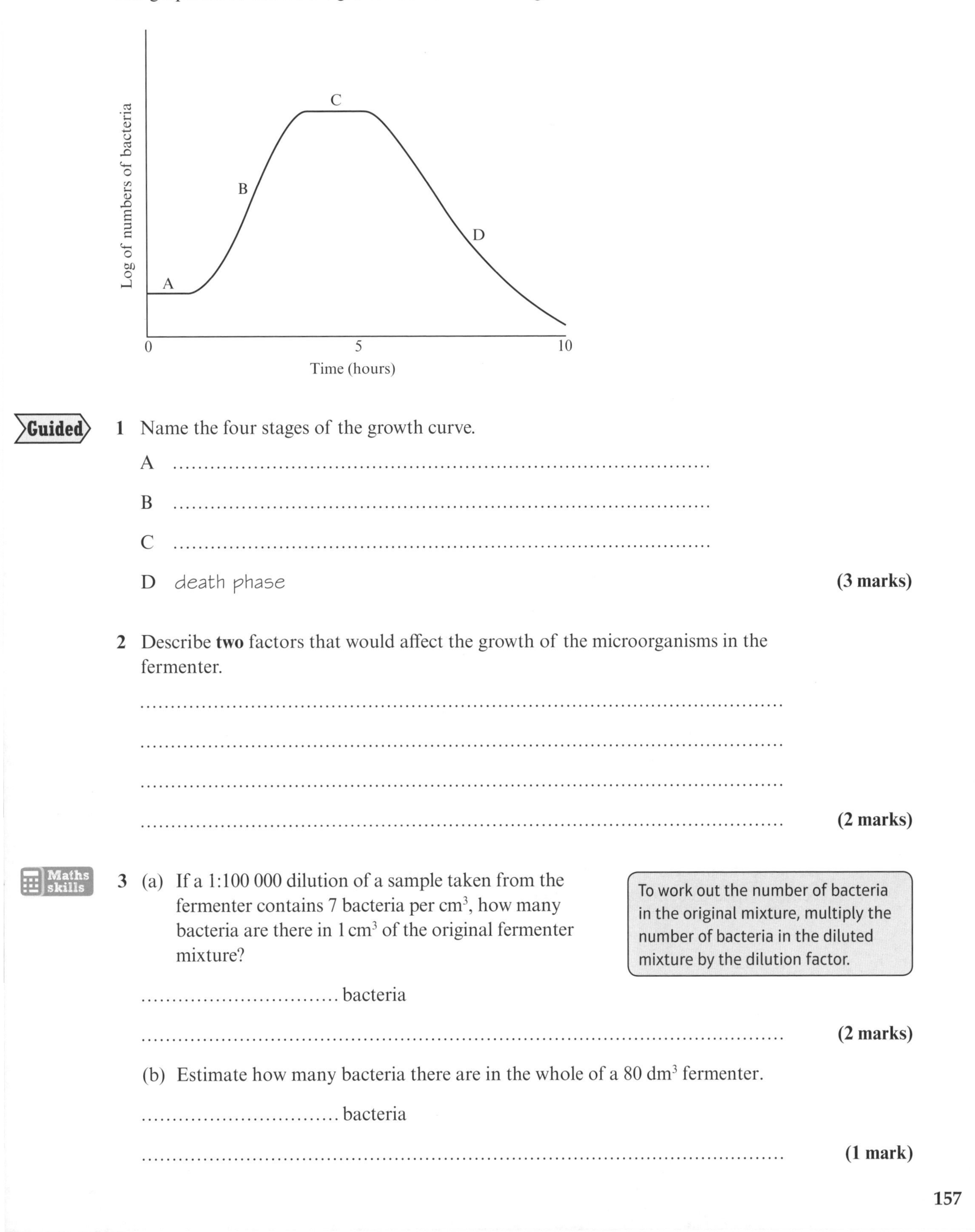

Guided

1 Name the four stages of the growth curve.

A ....

B ....

C ....

D *death phase* **(3 marks)**

2 Describe **two** factors that would affect the growth of the microorganisms in the fermenter.

....

....

....

.... **(2 marks)**

Maths skills

3 (a) If a 1:100 000 dilution of a sample taken from the fermenter contains 7 bacteria per cm^3, how many bacteria are there in 1 cm^3 of the original fermenter mixture?

> To work out the number of bacteria in the original mixture, multiply the number of bacteria in the diluted mixture by the dilution factor.

.... bacteria

.... **(2 marks)**

(b) Estimate how many bacteria there are in the whole of a 80 dm^3 fermenter.

.... bacteria

.... **(1 mark)**

Immobilised enzymes

Immobilised enzymes are used in industry for many different reactions.

1 Give **two** uses of immobilised enzymes.

....

....

....

.... **(2 marks)**

2 Describe **two** methods of immobilising enzymes.

....

....

....

....

....

....

....

.... **(4 marks)**

3 Explain the advantages of using immobilised enzymes.

Think about what could happen to enzymes if they were free in solution.

....

....

....

....

....

.... **(3 marks)**

4 Describe the process when the enzyme glucose isomerase, trapped inside alginate beads, is used to convert glucose to fructose.

....

....

....

.... **(2 marks)**

Exam skills

1 (a) Give **two** examples of how biotechnology is used in food production.

1 ....

2 .... **(2 marks)**

(b) Genetically modified *E. coli* are used in the production of human insulin in fermenters. The insulin is produced as a continuous culture.

(i) Describe the conditions needed in a fermenter in order to keep production at the optimum level.

....

....

....

....

....

.... **(3 marks)**

(ii) What are the advantages of producing insulin as a continuous culture?

A continuous culture is when reactants are continually added to the fermenter and the product is continually harvested.

....

....

....

....

....

.... **(2 marks)**

Practical skills

(iii) Explain any precautions that must be carried out when using a fermenter.

....

....

....

.... **(2 marks)**

Ecosystems

1 Ecosystems contain biotic and abiotic factors.

(a) What are **abiotic factors**?

....

.... **(1 mark)**

(b) List **two** abiotic factors.

....

.... **(2 marks)**

(c) Interspecific competition is a biotic factor. What is **interspecific competition**?

.... **(1 mark)**

2 A large tree ecosystem consists of the tree and several bird and insect species.

(a) Name **one** biotic factor affecting this ecosystem.

.... **(1 mark)**

(b) Suggest what would happen to population of the bird species if the insect populations decreased in number.

....

....

....

.... **(2 marks)**

(c) One summer is particularly hot. Explain what you think will happen to the ecosystem. Justify your answer.

Present a reasoned case for your answer.

....

....

....

....

....

....

....

.... **(3 marks)**

Biomass transfer

1 (a) What is meant by the term **gross primary productivity** (GPP)?

....

.... **(1 mark)**

(b) List **two** ways in which energy is lost from a food chain.

1 ....

2 .... **(2 marks)**

(c) What is a **trophic level**?

....

.... **(1 mark)**

(d) Explain why there are not usually more than five trophic levels in an ecosystem.

....

.... **(1 mark)**

2 (a) Plan a method to measure the energy content of a field of cabbages.

....

....

....

....

....

.... **(4 marks)**

(b) Evaluate any problems with using this method to estimate the energy content of a trophic level.

....

....

.... **(2 marks)**

Nitrogen cycle

1 Describe how nitrogen in the atmosphere cycles through plants and back into the atmosphere.

Include the names of bacteria involved in these processes.

....

(5 marks)

2 (a) Name the type of plant that has nodules that contain nitrogen-fixing bacteria.

....

(1 mark)

(b) Explain how this type of plant has a symbiotic relationship with nitrogen-fixing bacteria.

....

(2 marks)

Guided

3 Suggest how farmers can increase the levels of organic nitrogen in the soil.

Add fertilisers to the soil. ....

(2 marks)

Carbon cycle

1 (a) Describe how microorganisms can contribute to carbon dioxide being released into the atmosphere.

....

....

....

.... **(2 marks)**

Guided

(b) Explain how carbon dioxide can be stored away from the atmosphere by natural processes.

Some of the carbon dioxide is stored in plants by the process of

photosynthesis. ....

....

....

....

.... **(3 marks)**

2 In recent times, the concentration of carbon dioxide in the atmosphere has increased.

(a) Describe and explain the effect of the increase in carbon dioxide concentration on the Earth's temperature.

....

....

....

.... **(2 marks)**

(b) Suggest how the increase in atmospheric carbon dioxide could affect animals.

Think about the effect on habitat and food sources.

....

....

....

....

....

.... **(3 marks)**

Primary succession

1 Outline the process of primary succession.

....

....

....

....

....

....

....

.... **(4 marks)**

2 Name **two** examples of pioneer species on a glacier.

....

.... **(2 marks)**

3 One part of a sand dune is covered with hare's foot clover and bird's foot trefoil.

(a) The local council removes the plants from the sand dune to return the dune to bare sand. What is this process called?

.... **(1 mark)**

(b) Explain, with examples, any adaptation pioneer species on the sand dune would need.

Think about the adaptations of xerophyte species.

....

....

....

....

....

....

....

.... **(4 marks)**

Sampling

1 (a) What is **sampling**?

....

.... **(1 mark)**

Practical skills

(b) Describe how an area of heathland could be sampled using an interrupted line transect.

....

....

....

.... **(2 marks)**

(c) Name **two** abiotic factors that should be recorded along the line transect.

....

.... **(2 marks)**

2 The abundance of red squirrels in an area of woodland was measured using the capture, release and recapture method.

Guided

(a) What should be assumed about the population of the red squirrels in this area when using this method?

Marked and unmarked animals have an equal chance of being captured.

....

....

....

.... **(3 marks)**

Maths skills

(b) Use the formula below to estimate the population of red squirrels if 24 squirrels were caught and tagged the first time, and 27 were caught the second time, and 8 of these had been tagged in the first capture.

$$\text{population estimate} = \frac{\text{total first caught and marked} \times \text{total caught second time}}{\text{number of marked organisms in second catch}}$$

....

....

....

.... **(2 marks)**

Population sizes

*Read the following three statements:

Make sure you structure your answer logically, showing how the points you make are related or follow on from each other where appropriate. You should also support your points with relevant biological facts and/or evidence

- A species of finch lives in a tree habitat and feeds exclusively on tree spiders.
- Another species of finch moves into the same tree habitat and also feeds exclusively on tree spiders.
- The first species of finch produces an average of three eggs per nest and has one nest per year, whereas the second species of finch produces four eggs per nest and has two nests per year.

Describe and explain the population growth of both finch species and the tree spiders. Use the information given above and your knowledge of predator prey relationships.

....

....

....

....

....

....

....

....

....

....

....

....

....

....

....

....

....

....

....

.... **(9 marks)**

Conservation

1 (a) What is **conservation**?

....

.... **(1 mark)**

(b) Give **one** economic reason and **one** ecological reason for conservation.

....

....

....

.... **(2 marks)**

(c) Explain what is meant by **sustainable conservation**.

....

....

....

.... **(2 marks)**

2 Timber production is often managed in a sustainable way.

Guided

(a) Compare and contrast selective and clear felling.

Selective felling removes some trees from an area. ....

....

....

....

....

....

.... **(3 marks)**

(b) An arrow-making company decides to use coppicing as a sustainable method of timber production. Suggest why this may be the best method for the company.

....

....

....

.... **(2 marks)**

Managing an ecosystem

1 Explain why some ecosystems need to be managed.

....

....

....

....

....

....

....

.... **(4 marks)**

2 Describe some methods of managing ecosystems.

....

....

....

.... **(2 marks)**

3 Give **two** examples of actively managed ecosystems.

1 ....

2 .... **(2 marks)**

4 Suggest some problems associated with managing an ecosystem where people live.

Think about the things that people need to do every day.

....

....

....

....

....

.... **(3 marks)**

Exam skills

1 A species of yeast, *Saccharomyces cerevisiae*, is added to a flask of nutrient broth.

(a) Sketch a graph to show the population growth over time.

(2 marks)

(b) Explain the shape of the graph you have sketched.

Explain what happens at each stage of the growth curve.

....

....

....

....

....

....

....

.... **(4 marks)**

(c) Another species of yeast, *S. bayanus*, is added to the flask.

(i) Suggest what you would expect to happen to the population size of *S. cerevisiae*.

....

....

....

.... **(2 marks)**

(ii) What type of competition is this?

.... **(1 mark)**

OCR publishes official Sample Assessment Material on its website. This timed test has been written to help you practise what you have learned and may not be representative of a real exam paper.

Remember that the official OCR specification and associated assessment guidance materials are the only authoritative source of information and should always be referred to for definitive guidance.

AS Paper 1

Time: 1 hour 30 minutes

SECTION A

You should spend a maximum of 20 minutes on this section.

Answer all the questions.

---

1 Microscopes are useful tools for studying organisms. Which of the following statements is/are true:

Statement 1: Resolution is limited by the wavelength of light and diffraction of light as it passes through samples.

Statement 2: The resolving power of a transmission electron microscope is higher than that of a scanning electron microscope.

Statement 3: Electron microscopy has increased magnification but not increased resolution when compared to light microscopy.

☐ **A** 1, 2 and 3
☐ **B** only 1 and 2
☐ **C** only 2 and 3
☐ **D** only 1 **(1)**

2 Translation takes place in or on the...

☐ **A** Golgi apparatus.
☐ **B** lysosome.
☐ **C** nucleus.
☐ **D** ribosome. **(1)**

3 Which of the following best describes the role of the smooth endoplasmic reticulum?

☐ **A** control of substances into and out of the cell
☐ **B** synthesis and storage of lipids and carbohydrates
☐ **C** modification of proteins
☐ **D** synthesis and transport of proteins **(1)**

4 A triplet of bases that could not be found in mRNA is...

- ☐ **A** adenine adenine guanine.
- ☐ **B** adenine thymine guanine.
- ☐ **C** adenine cytosine guanine.
- ☐ **D** adenine uracil guanine. **(1)**

5 An image of a cell measures 8 cm but has an actual diameter of 10 μm. What is the magnification?

- ☐ **A** ×800
- ☐ **B** ×1250
- ☐ **C** ×8000
- ☐ **D** ×12 500 **(1)**

6 mRNA is made in the...

- ☐ **A** nucleus.
- ☐ **B** ribosomes.
- ☐ **C** endoplasmic reticulum.
- ☐ **D** Golgi apparatus. **(1)**

7 What reagent should be added to a solution in order to detect the presence of reducing sugars?

- ☐ **A** biuret solution
- ☐ **B** ethanol
- ☐ **C** Benedict's solution
- ☐ **D** iodine **(1)**

8 The pathway by which water moves across the root but does not enter living cells is called the...

- ☐ **A** symplast.
- ☐ **B** tonoplast.
- ☐ **C** apoplast.
- ☐ **D** amyloplast. **(1)**

9 Catalase produces 25 cm^3 of oxygen from 2 M hydrogen peroxide (H_2O_2) and 30 cm^3 of oxygen from 4 M H_2O_2 in 5 minutes. What is the percentage increase in the amount of oxygen produced from 4 M H_2O_2 compared to 2 M H_2O_2?

- ☐ **A** 5%
- ☐ **B** 10%
- ☐ **C** 20%
- ☐ **D** 25% **(1)**

10 The cellular tissue in the root into which all water must pass, due to the Casparian strip, is called the...

- ☐ **A** epidermis.
- ☐ **B** pericycle.
- ☐ **C** endodermis.
- ☐ **D** cortex. **(1)**

11 Which row best identifies the tissues responsible for the following roles in plants?

	Transport of water	Transport of assimilates	Cell division	Controlling water movement
☐ **A**	xylem	phloem	endodermis	cambium
☐ **B**	phloem	phloem	cambium	endodermis
☐ **C**	xylem	phloem	cambium	endodermis
☐ **D**	xylem	xylem	endodermis	cambium

(1)

12 Water moves into root hair cells by...

- ☐ **A** osmosis.
- ☐ **B** active transport.
- ☐ **C** facilitated diffusion.
- ☐ **D** endocytosis. **(1)**

13 Which of the following statements are true about arteries?

Statement 1: They contain elastic fibres to enable them to withstand the force of blood pumped out of the heart.

Statement 2: They contain valves to prevent backflow of blood.

Statement 3: They contain smooth muscle to enable them to withstand the force of blood pumped out of the heart.

- ☐ **A** 1, 2 and 3
- ☐ **B** only 1 and 3
- ☐ **C** only 2 and 3
- ☐ **D** only 1 **(1)**

14 Which row correctly describes the conditions required for tissue fluid formation?

	Water potential/kPa		Hydrostatic pressure/kPa		
	Capillary	Tissue	Arteriole end of capillary	Venule end of capillary	Tissue
☐ **A**	−1.3	−3.3	4.3	1.6	1.1
☐ **B**	−1.3	−3.3	1.6	4.3	1.1
☐ **C**	−3.3	−1.3	1.6	4.3	1.1
☐ **D**	−3.3	−1.3	4.3	1.6	1.1

(1)

15 A protein sample was separated by chromatography and was found to have two components. If the solvent front was 14 cm and the top component moved 9.5 cm, what is the R_f value of the top component?

- ☐ **A** 0.45
- ☐ **B** 1.47
- ☐ **C** 0.13
- ☐ **D** 0.68

(1)

16 During DNA replication complementary base pairing takes place. Which row best describes this process?

	Base pairing	Bonds formed	Enzyme involved
☐ **A**	A–C, T–G	hydrogen	DNA helicase
☐ **B**	A–T, C–G	hydrogen	DNA polymerase
☐ **C**	A–C, T–G	hydrogen	DNA polymerase
☐ **D**	A–T, C–G	ester	DNA polymerase

(1)

17 What happens during the anaphase stage of mitosis?

- ☐ **A** The sister chromatids are separated to the opposite sides of the cell.
- ☐ **B** The chromosomes form pairs in the centre of the cell.
- ☐ **C** The homologous chromosomes are separated to the opposite sides of the cell.
- ☐ **D** The chromosomes line up in the centre of the cell.

(1)

18 Which of the following terms means a cell that can become any type of body cell?

- ☐ **A** multipotent
- ☐ **B** differentiate
- ☐ **C** pluripotent
- ☐ **D** totipotent

(1)

19 Which of the following is characteristic of a non-competitive inhibitor?

- ☐ **A** It binds to an allosteric site on the enzyme.
- ☐ **B** It is usually reversible.
- ☐ **C** It binds to the active site on the enzyme.
- ☐ **D** The inhibitory effect can be overcome by increasing the substrate concentration.

(1)

20 Which of the following is the test for non-reducing sugars?

- ☐ **A** Add Benedict's reagent, then dilute acid and heat.
- ☐ **B** Add dilute acid and heat, then add Benedict's reagent and heat.
- ☐ **C** Add hydrogen sodium carbonate and heat, then add Benedict's reagent and heat.
- ☐ **D** Add Benedict's reagent and heat, then add dilute acid and heat.

(1)

SECTION B

Answer all the questions.

---

21 The figure shows a β-glucose molecule, which is the monomer of cellulose.

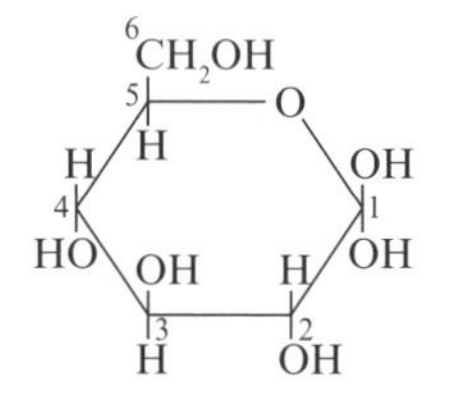

(a) Draw the products that are formed from a condensation reaction between two β-glucose molecules. **(2)**

(b) Explain how the structure of cellulose makes it a suitable molecule to form the walls of cells. **(4)**

A study was carried out in which thin films of cellulose were put in contact with an enzyme, cellulase, which breaks cellulose down. The films were weighed at intervals and the loss in mass, due to cellulose digestion, was recorded. This was done at five pH values: 3, 4, 5, 7 and 10.

The data are shown in the graph.

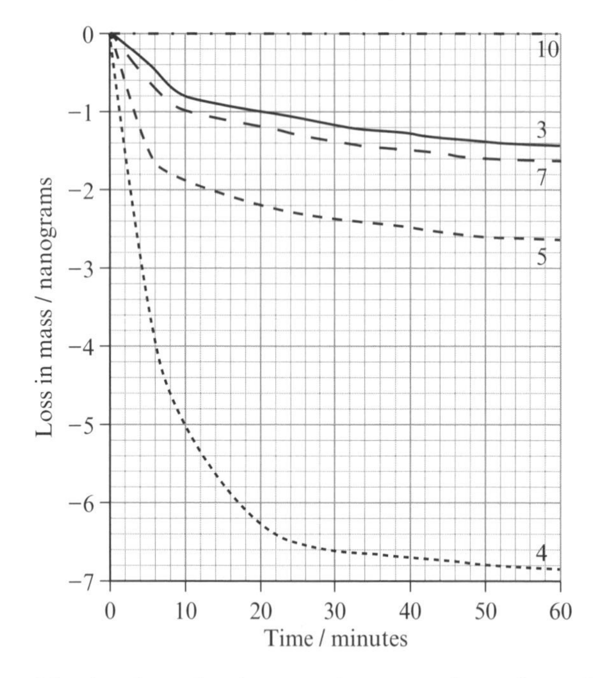

(c) Analyse the data to plot a graph to show the initial rate of reaction of this enzyme. **(4)**

(Total for Question 21 = 10 marks)

22 Genes are made up of DNA sequences that must be transcribed and translated to make proteins.

(a) What is the messenger RNA sequence for this DNA sequence?

TAGCCTAAACGCATG **(1)**

(b) Where does transcription take place? **(1)**

(c) Explain what a codon is and how it is used during translation. **(3)**

(d) Below is a figure of RNA codons. What is the amino acid sequence for the messenger RNA sequence in (a)? **(2)**

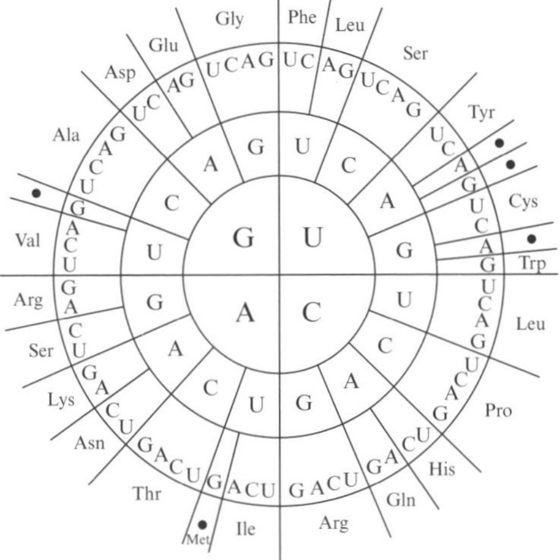

(e) A cell synthesises some proteins for export from the cell. Describe the role of organelles in the export of these proteins. **(3)**

(Total for Question 22 = 10 marks)

23 The figure shows part of the insect gas exchange system.

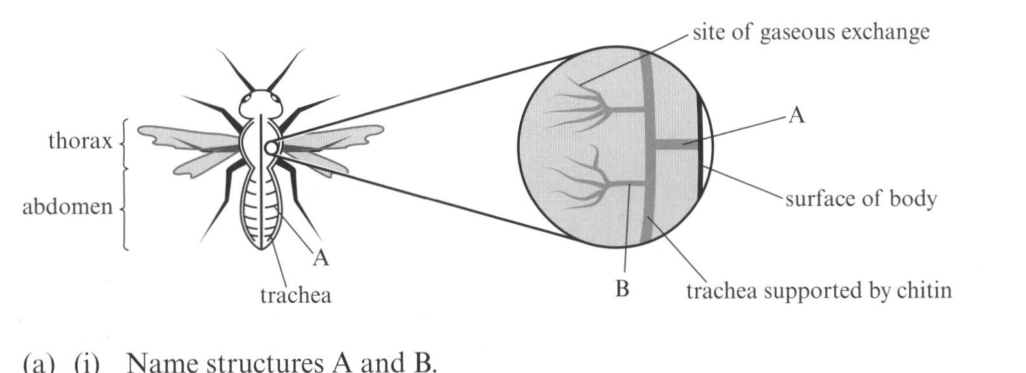

(a) (i) Name structures A and B. **(2)**

(ii) How do insects ensure the maximum amount of gaseous exchange occurs when they are highly active? **(2)**

(iii) Explain one way in which some insects can further ventilate their tracheal system using body movements. **(2)**

(b) The human gaseous exchange system is vastly different from that of insects, and most gas exchange occurs at the alveoli.

(i) State the name of the specialised tissue that constitutes the alveoli walls. **(1)**

(ii) What features of the alveoli enable efficient gas exchange? **(4)**

(Total for Question 23 = 11 marks)

24 Plasma membranes are a fluid mosaic made of phospholipids, proteins, glycoproteins and cholesterol.

(a) Complete the following passage about movement across the plasma membrane.

Small, non-polar molecules can cross the plasma membrane by

.... Small, polar molecules can cross the plasma

membrane through .... proteins by facilitated diffusion.

Large molecules can cross the plasma membrane by ....

This is an .... process and requires .... **(5)**

(b) Slices of beetroot were placed into test tubes of water at different temperatures. The amount of beetroot pigment in the surrounding water was estimated using pre-prepared colour standards. The data are shown below.

Temperature/°C	Percentage of pigment in the surrounding water/%
10	1
20	2
30	5
40	15
50	60
60	80

(i) Describe the effect of temperature on the percentage of pigment in the water. **(2)**

(ii) Explain how the pigment crossed the plasma membrane into the water. **(3)**

(iii) Suggest a limitation with this study. **(1)**

(Total for Question 24 = 11 marks)

25 A factory is suspected of polluting a local river with nitrates. Several river samples have been taken along the river and brought back to the laboratory to be analysed by a colorimeter.

(a) Describe how a colorimeter is used to estimate the concentration of nitrates. **(4)**

(b) Use the calibration curve and table below to work out the concentration of nitrates in the samples. **(3)**

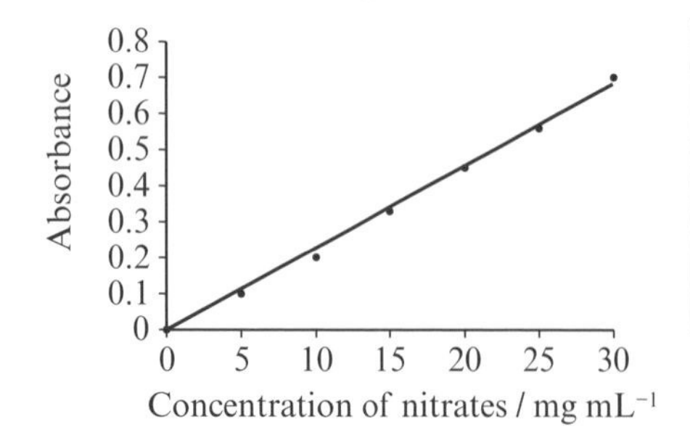

Sample	Absorbance	Concentration of nitrates/mg mL^{-1}
A	0.1	
B	0.4	
C	0.2	

(c) Sample A was taken upriver from the factory, sample B from close to the factory, and sample C downstream from the factory. Justify whether it is possible to conclude that the river is being polluted by the factory. **(2)**

(Total for Question 25 = 9 marks)

OCR publishes official Sample Assessment Material on its website. This timed test has been written to help you practise what you have learned and may not be representative of a real exam paper.

Remember that the official OCR specification and associated assessment guidance materials are the only authoritative source of information and should always be referred to for definitive guidance.

AS Paper 2

Time: 1 hour 30 minutes

Answer all the questions.

---

1 Proteins are made out of long chains of amino acids.

(a) What is the name of the bond between amino acids? **(1)**

(b) (i) Draw a figure to show how two amino acids join together to form a dipeptide. **(3)**

(ii) What is this reaction called? **(1)**

(c) Outline the processes that lead to polypeptides folding into tertiary structures. **(4)**

(Total for Question 1 = 9 marks)

2 Potato slices were placed into salt solutions of different concentrations. The mass of the potato slices was measured before and after being placed into the salt solution.

(a) The graph below shows the percentage change in mass of the potato slices after 12 hours.

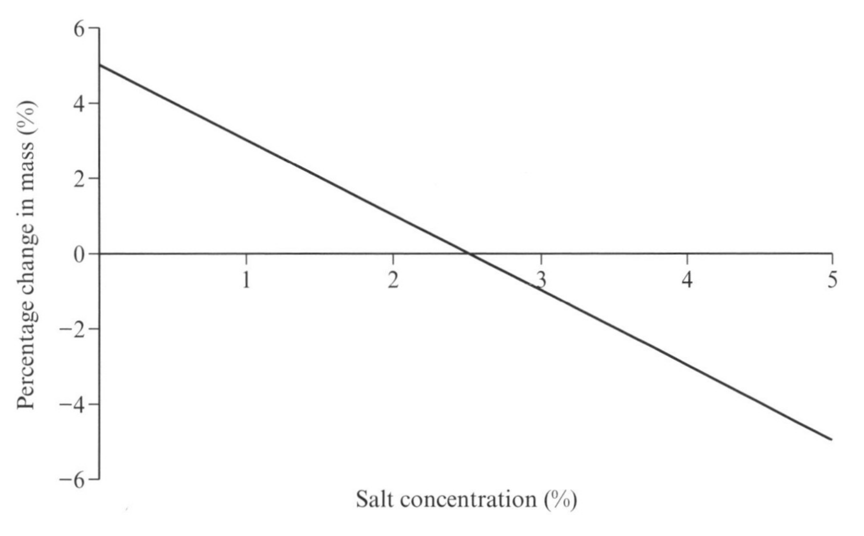

(i) What is the percentage change in mass of the potato slice in 3% salt solution? **(2)**

(ii) Use the graph to find out and explain when the water potential of the potato slice is the same as the salt solution. **(3)**

(b) At 0% salt concentration, the percentage change in mass of the potato slice is 5%.

(i) What evidence from the graph shows that water is lost from the potato cells at 4% salt concentration? **(1)**

(ii) Explain what happens to the cells of the potato slice when water is lost. **(2)**

(c) A student wanted to investigate if the same changes in mass would be measured if sugar solutions were used instead of salt solutions.

This was the method used:

1. Potato slices were placed into sugar solutions of 0%, 2%, 4%, 8%, and 10% concentration.
2. After 12 hours the mass of the potato slices was measured using a weighing balance.

(i) What is the independent variable in this investigation? **(1)**

(ii) Suggest some variables that should be kept constant in this investigation. **(4)**

(iii) Refine the method to make it more valid and reliable. **(4)**

(Total for Question 2 = 17 marks)

3 A spirometer is used to measure the air capacity of the lungs.

(a) The graph below shows the lung capacity of a student athlete.

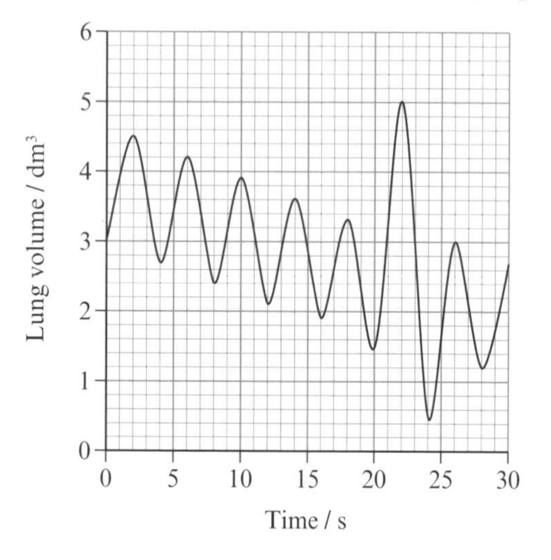

(i) What is meant by the vital capacity? **(1)**

(ii) Use the graph to measure the student's vital capacity. **(2)**

(b) The graph can be used to calculate the student's breathing rate and oxygen consumption.

(i) What is the student's breathing rate? Show your working. **(2)**

(ii) The student uses 1.5 dm^3 of oxygen in 20 seconds. Calculate the student's oxygen consumption rate. Show your working. **(2)**

(c) The spirometer measured the vital capacity of another student as 3.4 dm^3.

(i) Suggest why the vital capacity of the second student was lower than that of the first student. **(2)**

(ii) What precautions should be taken before allowing another student to use the spirometer? **(2)**

(Total for Question 3 = 11 marks)

4 Tuberculosis (TB) is a disease that can easily spread from person to person.

(a) The graph below shows the number of TB deaths in males and females in the UK from 1911 to 2005.

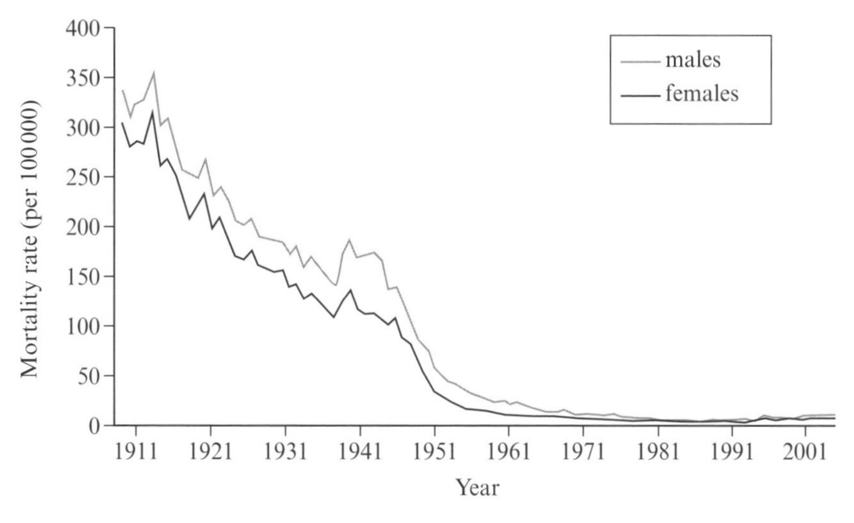

(i) What type of microorganism causes TB? **(1)**

(ii) Suggest what has caused the decline in deaths from TB since 1945. **(2)**

(b) (i) Describe how the cells of the immune system fight off the TB infection. **(5)**

(ii) What type of immunity is this an example of? **(1)**

(c) Discuss the implications of not giving the BCG vaccine to all teenagers in the UK. **(4)**

(Total marks for Question 4 = 13 marks)

5 The wolf, *Canis lupus*, is closely related to the coyote, *Canis latrans*.

(a) Complete the table below to show the classification of the wolf. **(6)**

Domain	
Kingdom	Animalia
	Chordata
Class	Mammalia
Order	Carnivora
	Canidae
Species	

(b) Describe the characteristics of a species belonging to the Animalia kingdom. **(3)**

(c) Suggest why wolves and coyotes do not occupy the same territory. **(3)**

(Total for Question 5 = 12 marks)

6 The table below shows the proportion of DNA nucleotides in four different species.

	Proportion of DNA nucleotides/%			
Species	**A**	**C**	**G**	**T**
Mycobacterium tuberculosis	16	36	34	14
sea urchin	30	17	19	34
mangrove tree	16	35	29	20
herring	27	20	22	31

(a) (i) Describe the patterns in the data. **(3)**

(ii) What evidence from the table shows how the nucleotides bind together? **(1)**

(b) DNA strands can be separated by applying heat.

(i) Which species' DNA would require the most heat to separate the DNA strands? **(1)**

(ii) Explain your answer to (i). **(3)**

(Total for Question 6 = 8 marks)

OCR publishes official Sample Assessment Material on its website. This timed test has been written to help you practise what you have learned and may not be representative of a real exam paper.

Remember that the official OCR specification and associated assessment guidance materials are the only authoritative source of information and should always be referred to for definitive guidance.

A Paper 1

Time: 2 hours and 15 minutes

SECTION A

You should spend a maximum of 20 minutes on this section.

Answer all the questions.

---

1 Which of the following best describes the role of the smooth endoplasmic reticulum?

☐ **A** control of substances into and out of the cell
☐ **B** synthesis and storage of lipids and carbohydrates
☐ **C** modification of proteins
☐ **D** synthesis and transport of proteins **(1)**

2 Which of the following is the ultimate cause for transmission of an action potential?

☐ **A** sodium ions and potassium ions
☐ **B** sodium/potassium ion pumps
☐ **C** sodium ions and voltage-gated sodium channels
☐ **D** the myelin sheath **(1)**

3 Which of the following will happen if an animal cell is placed in a solution with a lower water potential than the cell content?

☐ **A** Water will enter the cell and the cell will burst.
☐ **B** Sugar will enter the cell.
☐ **C** The volume of the cell will reduce and the cell will undergo crenation.
☐ **D** The cell will become turgid. **(1)**

4 Which row correctly describes the conditions required for tissue fluid formation?

	Water potential/kPa		Hydrostatic pressure/kPa		
	Capillary	**Tissue**	**Arteriole end of capillary**	**Venule end of capillary**	**Tissue**
☐ **A**	−1.3	−3.3	4.3	1.6	1.1
☐ **B**	−1.3	−3.3	1.6	4.3	1.1
☐ **C**	−3.3	−1.3	1.6	4.3	1.1
☐ **D**	−3.3	−1.3	4.3	1.6	1.1

(1)

5 Which of the following is not an example of a catabolic reaction?

☐ **A** synthesis of spindle microtubules during mitosis

☐ **B** formation of insulin in cells of the pancreas

☐ **C** conversion of glycogen to glucose in liver cells

☐ **D** formation of citrate in the Krebs cycle

6 During DNA replication complementary base pairing takes place. Which row best describes this process?

	Base pairing	Bonds formed	Enzyme involved
☐ **A**	A–C, T–G	hydrogen	DNA helicase
☐ **B**	A–T, C–G	hydrogen	DNA polymerase
☐ **C**	A–C, T–G	hydrogen	DNA polymerase
☐ **D**	A–T, C–G	ester	DNA polymerase

(1)

7 How does antidiuretic hormone affect the kidney?

☐ **A** decreases permeability of collecting ducts to water

☐ **B** increases permeability of collecting ducts to water

☐ **C** increases volume of blood passing into the nephron

☐ **D** increases concentration of salts in the medulla

(1)

8 Which row best identifies the tissues responsible for the following roles in plants?

	Transport of water	Transport of assimilates	Cell division	Controlling water movement
☐ **A**	xylem	phloem	endodermis	cambium
☐ **B**	phloem	phloem	cambium	endodermis
☐ **C**	xylem	phloem	cambium	endodermis
☐ **D**	xylem	xylem	endodermis	cambium

9 When maltose is hydrolysed it forms two simple sugars. Which figure shows the correct structures for products formed in this reaction? **(1)**

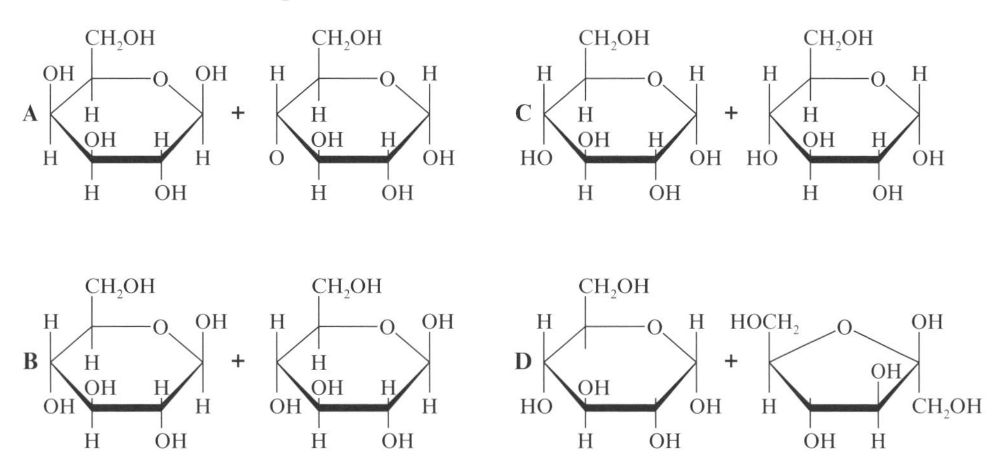

10 Which row correctly describes the responses of a plant to its environment?

	Response	Stimulus	Type of response
☐ **A**	A new root grows down into the soil.	gravity	geotropism
☐ **B**	A shoot grows upwards towards light.	water potential	phototropism
☐ **C**	A pollen tube grows towards the ovary.	gravity	thigmotropism
☐ **D**	A bean shoot curls around a stick.	chemical	chemotropism

(1)

11 Which row correctly describes the filaments in a muscle fibre and their action during a muscular contraction?

	Element	Made of	Attaches to	Action
☐ **A**	thick filament	myosin	actin	gets shorter
☐ **B**	thin filament	actin	myosin	move closer together
☐ **C**	thick filament	actin	myosin	move closer together
☐ **D**	thin filament	myosin	tropomyosin	gets shorter

(1)

12 Which of the following statements is/are true about photosynthesis?

Statement 1: Cyclic phosphorylation uses only photosystem I.

Statement 2: Both photosystem I and photosystem II use chlorophyll a.

Statement 3: Only photosystem I produces reduced NADP.

☐ **A** 1, 2 and 3

☐ **B** only 1 and 2

☐ **C** only 2 and 3

☐ **D** only 1 **(1)**

13 Which of the following statements is/are true about arteries?

Statement 1: They contain elastic fibres to enable them to withstand the force of blood pumped out of the heart.

Statement 2: They contain many valves to prevent backflow of blood.

Statement 3: They contain smooth muscle to control the blood flow.

☐ **A** 1, 2 and 3

☐ **B** only 1 and 3

☐ **C** only 2 and 3

☐ **D** only 1 **(1)**

14 Microscopes are useful tools for studying organisms. Which of the following statements is/are true?

Statement 1: Resolution is limited by the wavelength of light and diffraction of light as it passes through samples.

Statement 2: The resolving power of a transmission electron microscope is higher than that of a scanning electron microscope.

Statement 3: Electron microscopy has increased magnification but not increased resolution when compared to light microscopy.

☐ **A** 1, 2 and 3

☐ **B** only 1 and 2

☐ **C** only 2 and 3

☐ **D** only 1 **(1)**

15 Which of the following statements is/are true?

Statement 1: Enzymes catalyse metabolic reactions.

Statement 2: All enzyme inhibitors bind to the active site of the enzyme.

Statement 3: Enzymes work optimally at 37 °C.

☐ **A** 1, 2 and 3

☐ **B** only 1 and 2

☐ **C** only 2 and 3

☐ **D** only 1 **(1)**

SECTION B

Answer all the questions.

---

16 An investigation into photosynthesis and respiration in a leaf was carried out. The net uptake of carbon dioxide by the leaf in bright light and the mass of carbon dioxide released in the dark were measured at different temperatures. The results are shown in the table.

	Temperature/°C				
	10	**20**	**30**	**40**	**50**
net uptake of carbon dioxide in bright light (mg g^{-1} dry mass $hour^{-1}$)	2.4	3.9	2.2	0.3	0.0
release of carbon dioxide in the dark (mg g^{-1} dry mass $hour^{-1}$)	0.7	1.4	3.2	6.1	0.0
true rate of photosynthesis (mg g^{-1} dry mass $hour^{-1}$)					

(a) Assuming that the rate of release of carbon dioxide is the same in daylight as in the dark, calculate the true rate of photosynthesis at each temperature. **(2)**

(b) Calculate the temperature coefficients (Q10) for respiration and photosynthesis between 20 °C and 30 °C. Express your answer to 1 decimal place. **(3)**

(c) Comment on the Q10 values for respiration and photosynthesis at this temperature range. **(4)**

The figure shows the result of an investigation where the rate of photosynthesis was determined in increasing light intensities under differing temperatures and CO_2 concentrations.

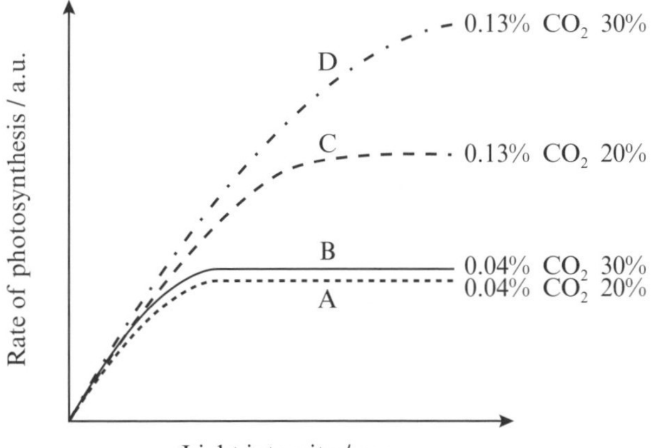

Experiments A and B both reached the maximum rate of photosynthesis when the light source was 27 cm from the leaf. The following equation was used to determine the light intensities used in this experiment:

$I = \frac{1}{d^2}$, where I = light intensity and d = distance (m)

(d) State the light intensity at which both A and B reached the maximum rate of photosynthesis to 3 significant figures? **(2)**

(e) Using the figure, state the limiting factor, other than light intensity, that has the greatest effect on the rate of photosynthesis according to this experiment. **(1)**

(f) Which experiment (A, B, C or D) shows the rate of photosynthesis when temperature is the main limiting factor? **(1)**

(g) Sketch a line on the graph to indicate what you would expect to happen to the rate of photosynthesis if the CO_2 concentration was reduced to 0.01% at 20 °C. **(1)**

(h) Explain the effect of reduced CO_2 concentration on the rate of photosynthesis. You may use a figure to help with your answer. **(6)**

(Total for Question 16 = 20 marks)

17 The figure shows an outline of an important metabolic process carried out in the liver.

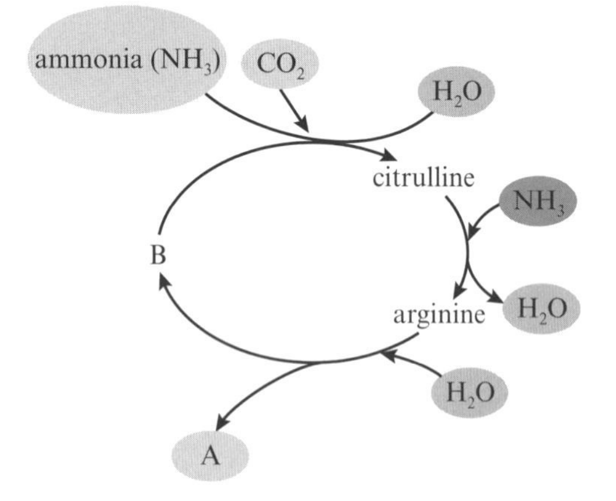

(a) State the name given to this process. **(1)**

(b) With reference to the figure, name A and B. **(2)**

(c) How is ammonia formed before this process can take place? **(1)**

(d) Why is ammonia converted in the body into substance A rather than being excreted? **(2)**

(e) The liver is also responsible for the detoxification of many compounds, including alcohol. Outline how the detoxification of alcohol can lead to the release of energy. **(4)**

(Total for Question 17 = 10 marks)

18 (a) The figure shows part of the insect gas exchange system.

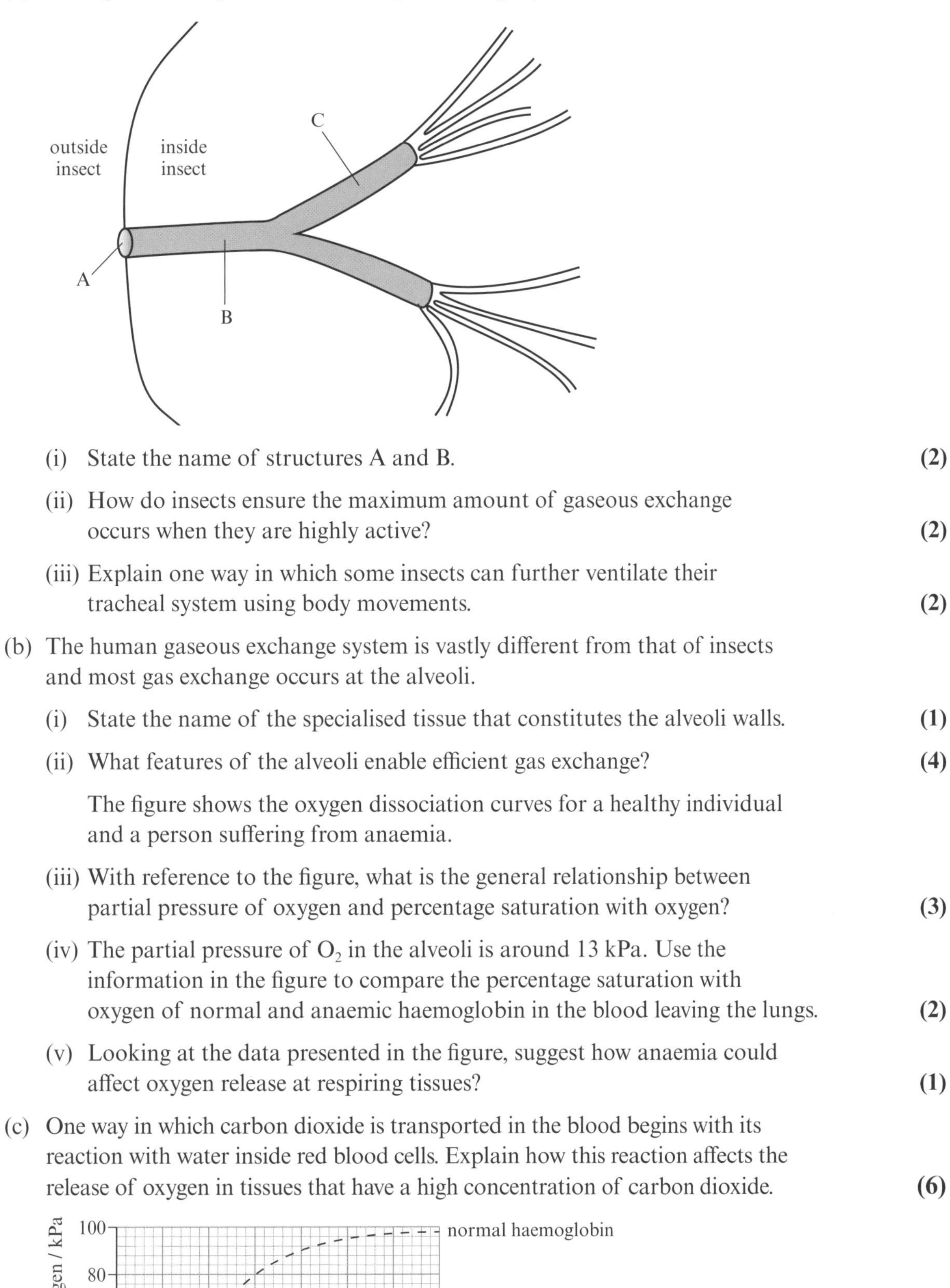

(i) State the name of structures A and B. **(2)**

(ii) How do insects ensure the maximum amount of gaseous exchange occurs when they are highly active? **(2)**

(iii) Explain one way in which some insects can further ventilate their tracheal system using body movements. **(2)**

(b) The human gaseous exchange system is vastly different from that of insects and most gas exchange occurs at the alveoli.

(i) State the name of the specialised tissue that constitutes the alveoli walls. **(1)**

(ii) What features of the alveoli enable efficient gas exchange? **(4)**

The figure shows the oxygen dissociation curves for a healthy individual and a person suffering from anaemia.

(iii) With reference to the figure, what is the general relationship between partial pressure of oxygen and percentage saturation with oxygen? **(3)**

(iv) The partial pressure of O_2 in the alveoli is around 13 kPa. Use the information in the figure to compare the percentage saturation with oxygen of normal and anaemic haemoglobin in the blood leaving the lungs. **(2)**

(v) Looking at the data presented in the figure, suggest how anaemia could affect oxygen release at respiring tissues? **(1)**

(c) One way in which carbon dioxide is transported in the blood begins with its reaction with water inside red blood cells. Explain how this reaction affects the release of oxygen in tissues that have a high concentration of carbon dioxide. **(6)**

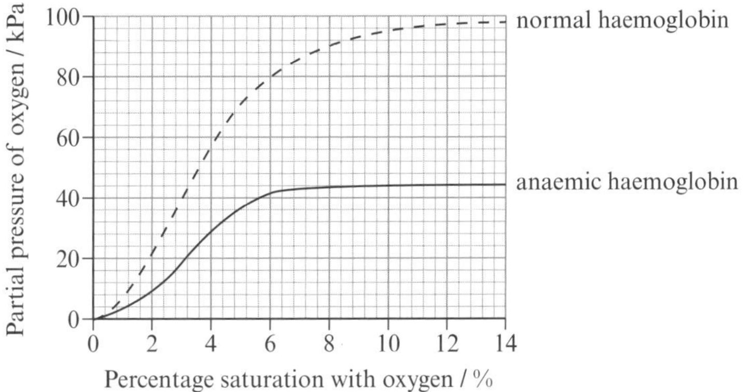

(Total for Question 18 = 23 marks)

19 A student investigated the effect of plant growth substances (IAA and GA) on the elongation of stems. The student marked a 10 mm length of stem on each of four plants and applied growth substances to the stem. The graph shows the student's results.

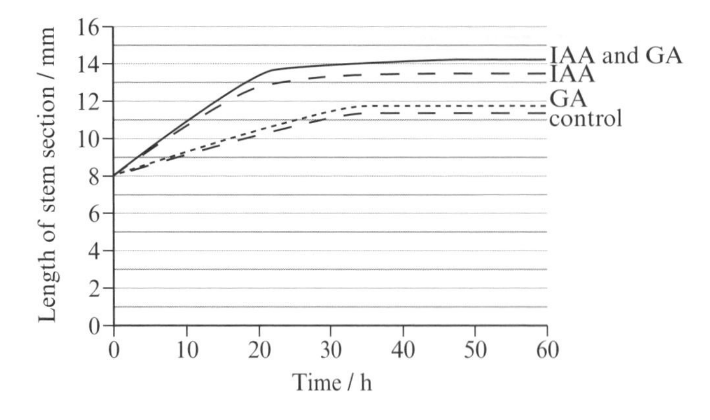

(a) Using the maximum growth of each stem, calculate the percentage increase in growth caused by IAA and GA together, compared to the control. **(2)**

(b) Suggest a suitable treatment for the control. **(1)**

(c) Identify two other variables that the student would need to control in this investigation if light intensity, temperature and humidity were kept constant. **(2)**

(d) Bananas produce a hormone that speeds up the ripening process.

(i) What is the name of this hormone? **(1)**

(ii) This hormone also promotes fruit drop in cherry trees. Suggest how this might be of an economic benefit to farmers. **(1)**

(Total for Question 19 = 7 marks)

20 (a) What are the main differences between type 1 and type 2 diabetes? **(3)**

(b) In patients with diabetes excess glucose can be detected in the urine. Suggest why this might be. **(2)**

(c) A student was concerned that she might be developing diabetes and performed a simple Benedict's test on her urine to determine its glucose concentration. The graph shows the calibration curve she produced.

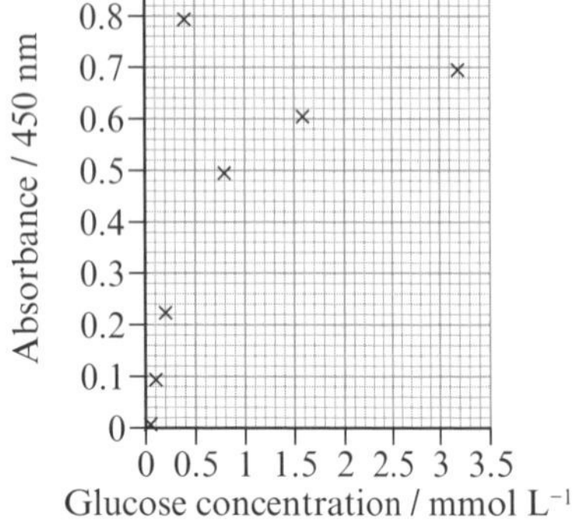

(i) Outline the procedure that the student performed to produce this calibration curve. **(3)**

(ii) The line of best fit is missing on the calibration curve. Draw the line of best fit. **(2)**

(iii) The student's urine sample gave an absorbance reading at 450 nm of 0.42. Using the graph and your line of best fit determine the glucose concentration of the student's urine. **(1)**

(iv) The normal range of glucose found in urine is between 0 and 0.8 mmol L^{-1}. What can the student conclude from her results. **(1)**

(v) Suggest two improvements that the student could include to increase the validity of her experiment? **(2)**

(vi) What alternative test could the student use to determine the glucose concentration in her urine and why might this test produce a more valid result? **(2)**

(Total for Question 20 = 16 marks)

OCR publishes official Sample Assessment Material on its website. This timed test has been written to help you practise what you have learned and may not be representative of a real exam paper.

Remember that the official OCR specification and associated assessment guidance materials are the only authoritative source of information and should always be referred to for definitive guidance.

A Paper 2

Time: 2 hours and 15 minutes

SECTION A

You should spend a maximum of 20 minutes on this section.

Answer all the questions.

---

1 An experiment was set up to test a sample for the presence of sucrose. Benedict's tests for reducing and non-reducing sugars were performed. Which row in the table represents the results that would be obtained?

	Colours observed after testing	
	Benedict's test for reducing sugars	**Benedict's test for non-reducing sugars**
☐ **A**	blue	blue
☐ **B**	brick-red	blue
☐ **C**	blue	brick-red
☐ **D**	brick-red	brick-red

(1)

2 Which of these is a feature of collagen?

☐ **A** three polypeptide chains held together by ionic bonds

☐ **B** every third amino acid is lysine

☐ **C** fibrous structure

☐ **D** globular structure

(1)

3 In the following graphs the arrow indicates re-exposure to a pathogen. Which of these graphs shows a primary and secondary immune response to the pathogen?

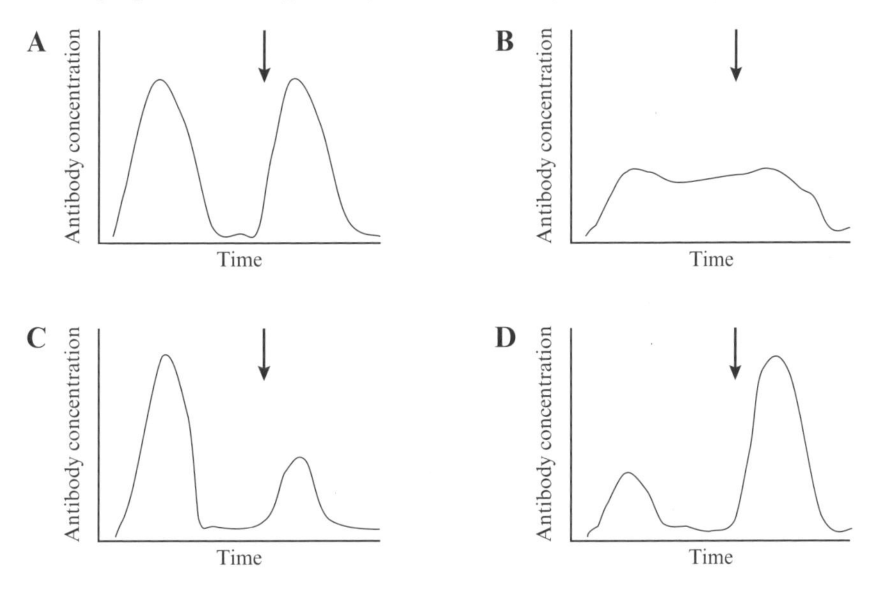

4 What is the name given to the signalling molecules involved in the immune response?

- ☐ **A** cytokines
- ☐ **B** antibodies
- ☐ **C** cytokinins
- ☐ **D** antigens **(1)**

5 Which of the following enzymes is involved in transcription?

- ☐ **A** DNA polymerase
- ☐ **B** RNA transcriptase
- ☐ **C** RNA polymerase
- ☐ **D** DNA helicase **(1)**

6 A scientist is setting up a culture plate of bacteria by first making serial dilutions as shown in the figure. What is the concentration of the bacteria in the original tube?

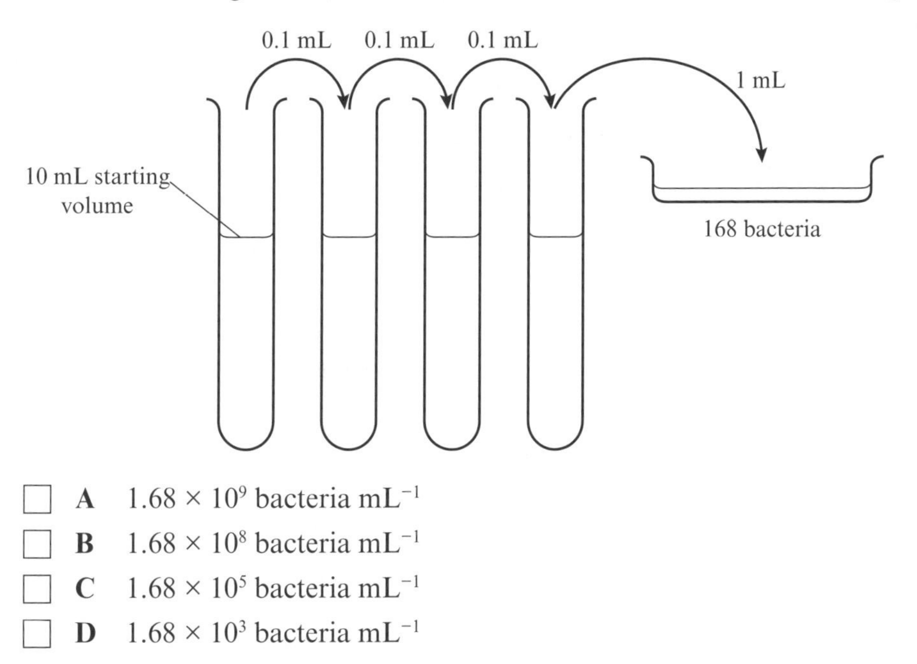

- ☐ **A** 1.68×10^9 bacteria mL^{-1}
- ☐ **B** 1.68×10^8 bacteria mL^{-1}
- ☐ **C** 1.68×10^5 bacteria mL^{-1}
- ☐ **D** 1.68×10^3 bacteria mL^{-1} **(1)**

7 Which of the following diseases is caused by a fungal infection?

- ☐ **A** AIDS
- ☐ **B** ringworm
- ☐ **C** tuberculosis
- ☐ **D** ring rot **(1)**

8 Which of the following statements best describes apoptosis?

- ☐ **A** caused only by pathogen infection
- ☐ **B** the cell increases in size and bursts
- ☐ **C** important in development of body plans
- ☐ **D** a form of cell movement **(1)**

9 DNA electrophoresis is a technique used in DNA fingerprinting and genetic engineering. The figure shows an electrophoresis gel. Which band (A, B, C or D) is the smallest fragment of DNA? (1)

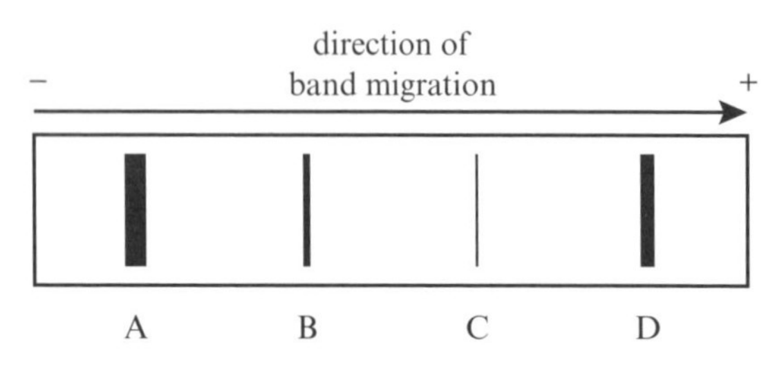

10 Which of the following statements best describes gene therapy?

- ☐ **A** transformation of a bacteria with a plasmid
- ☐ **B** producing insulin in goat's milk
- ☐ **C** inserting a correct copy of a defective gene into a human's cells
- ☐ **D** *in-vitro* fertilisation of a human egg with sperm from a donor **(1)**

11 In which of the following organisms would you find homeobox genes?

- ☐ **A** mouse
- ☐ **B** daisy
- ☐ **C** bread mould
- ☐ **D** all of the above **(1)**

12 Which of the following statements is/are true about seed banks?

Statement 1: They are a form of *ex-situ* conservation.

Statement 2: Seeds are kept in habitats like those of the parent plant.

Statement 3: Seeds are kept in a warm, humid environment.

- ☐ **A** 1, 2 and 3
- ☐ **B** only 1 and 2
- ☐ **C** only 2 and 3
- ☐ **D** only 1 **(1)**

13 Which of the following statements describes processes that occur during mitosis?

Statement 1: Sister chromatids are separated and move to opposite poles by the contraction of spindle fibres.

Statement 2: The nuclear envelope disintegrates during prophase.

Statement 3: Bivalents form where homologous chromosomes line up on the metaphase plate.

☐ **A** 1, 2 and 3
☐ **B** only 1 and 2
☐ **C** only 2 and 3
☐ **D** only 1 **(1)**

14 Which of the following statements is/are true about phylogenetics?

Statement 1: It is based on evolutionary relationships.

Statement 2: Extinct species can be classified by this method.

Statement 3: Classification is based on similarities in physical appearance.

☐ **A** 1, 2 and 3
☐ **B** only 1 and 2
☐ **C** only 2 and 3
☐ **D** only 1 **(1)**

15 Which of the following statements is/are true about epistasis?

Statement 1: The genes are always on the same chromosome.

Statement 2: The epistatic gene masks the expression of the hypostatic gene.

Statement 3: There is always more than one gene involved.

☐ **A** 1, 2 and 3
☐ **B** only 1 and 2
☐ **C** only 2 and 3
☐ **D** only 1 **(1)**

SECTION B

Answer all the questions.

---

16 Forest wildfires are spontaneous fires caused naturally in forest ecosystems where long periods of hot and dry weather occur. Although wildfires can cause billions of pounds worth of damage, many organisms have evolved to withstand them, and they are even essential to the life cycle of many plants. Since 2001 there have been, on average, 8000 wildfires per year in the US state of California.

(a) *Ceanothus* is a genus of shrubs whose seeds only germinate when temperatures reach 70–100 °C. How might this be an advantage in the forest ecosystems described above? **(2)**

(b) Explain the changes in plant species that occur following a forest wildfire until the appearance of mature trees. **(4)**

(c) After a forest fire it has been shown that levels of organic nitrogen in the soil can decrease by up to 50%. However, the levels of ammonium and nitrate often increase. Using your knowledge of the nitrogen cycle suggest why this could result in a quicker re-establishment of mature trees than deforestation. **(2)**

(Total for Question 16 = 8 marks)

17 The enzymes alcohol dehydrogenase (ADH) and acetaldehyde dehydrogenase (ALDH) are involved in the breakdown of ethanol to acetic acid. The level of activity of these enzymes differs in humans due to variations in the ADH and ALDH genes. The figure shows the metabolic pathway activated by these enzymes.

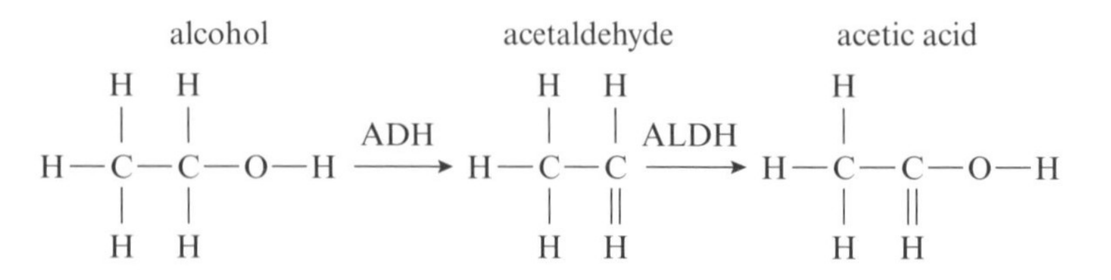

(a) Read the following four statements:

Statement 1: ADH4 is a gene coding for the alcohol dehydrogenase enzyme found in the stomach, oesophagus and tongue of primates. ADH4 is active against a wide range of alcohols.

Statement 2: Less effective ADH4 causes slow metabolism of alcohol and build-up of toxins, resulting in a lack of muscle coordination, poor judgement, nausea, vomiting, and memory lapse upon excessive alcohol consumption.

Statement 3: Just over 10 million years ago the Earth cooled, leading to a struggle for food sources. Fallen fruits provided an additional source of sustenance although breakdown by bacteria produces alcohol. This was also around the time that some primates became more terrestrial.

Statement 4: Alcohol consumption leads to increased risk-taking behaviour and increased aggression.

Use this information and your knowledge of Darwin's theory of evolution via natural selection to suggest why humans have an ADH4 that is at least 40 times more effective than that of lemurs and baboons. **(9)**

The ALDH2*1 allele is the most common ALDH allele found in humans worldwide. The ALDH2*2 allele is a dominant allele that is less common and is responsible for the increased adverse effects of alcohol consumption listed in statement 2, but decreased alcoholism.

Approximately 46% of people of East Asian descent have increased adverse effects of alcohol due to presence of the ALDH2*2 allele (heterozygotes and homozygotes), compared to 10% of people of Northern European descent.

(b) The Hardy–Weinberg principle states that:

$p + q = 1$ and $p^2 + 2pq + q^2 = 1$

where p is the frequency of the dominant allele and q is the frequency of the recessive allele in a population.

Using the information above, calculate the frequency of the ALDH2*2 allele in the East Asian population. Show your working. **(3)**

(c) The ALDH2*2 allele has arisen through a point mutation in the ALDH2 gene, resulting in a change in one amino acid. How could a change in one amino acid affect the activity of the enzyme? **(4)**

(d) Disulfiram is a drug used to treat alcoholism. It works by preventing the conversion of acetaldehyde to acetic acid. Using the information above and the figure above suggest why this drug may not be effective in people of East Asian descent. **(2)**

(Total for Question 17 = 18 marks)

18 Systemic lupus erythematosus (SLE or lupus) is a disease which is thought to be caused by the body producing antibodies against its own proteins – more specifically those that work in the nucleus of the cell.

(a) What is the scientific term for diseases such as SLE where the body produces self-antibodies? **(1)**

(b) Although lupus is often triggered by environmental factors, there is usually an underlying genetic defect. What molecular evidence could scientists look for to determine whether there is a genetic cause for an individual having lupus? **(2)**

(c) Studies have found that patients with SLE often have significantly fewer circulating T regulatory cells than people with the disease. Why could this cause a disease like SLE? **(2)**

(d) How would an antibody recognise a self-antigen? **(1)**

(e) Name another autoimmune disease other than lupus. **(1)**

(Total for Question 18 = 7 marks)

19 Malaria is a disease that is thought to have affected humans for at least 10 000 years.

(a) To what kingdom does *P. falciparum* belong? **(1)**

(b) Describe the primary non-specific defence mechanisms that humans have to fight off a *P. falciparum* infection. **(3)**

(c) If the malaria is not treated early enough an immune response is initiated.

(i) In the cell-mediated immune response, T lymphocytes are involved. Describe how the cells of the non-specific response and T lymphocytes interact. **(3)**

(ii) Humoral immunity involves the activation of B lymphocytes. How do B lymphocytes contribute to the immune response? **(3)**

(d) Once someone has contracted malaria the parasite remains undisturbed in the liver, re-emerging every 10 years or so. Why is it particularly important that a humoral immune response to malaria is initiated upon primary infection with the pathogen? **(2)**

(Total for Question 19 = 12 marks)

20 Berberine is a secondary metabolite produced by many plant species including the Californian poppy.

(a) A company called MediPop in California wanted to buy a plot of land to use for farming the poppy and extracting the berberine for its medicinal uses. In order to convert the land for industrial use the company initiated ecological surveys on two potential sites.

	Individuals per m²		
Species	**Site 1**	**Site 2**	$\left(\frac{n}{N}\right)^2$
U	8	17	
V	3	5	
W	0	13	
X	7	10	
Y	2	17	
Z	18	9	
Total			
D	0.64		

Simpson's index of biodiversity can be calculated using the following formula:

$$D = 1 - \left[\Sigma\left(\frac{n}{N}\right)^2\right]$$

where n is the number of a particular species and N is the total number of all species.

(i) Complete the table to calculate D for site 2. **(3)**

(ii) State which site the company should clear for growing the poppy and why? **(2)**

(iii) On closer examination of the sampling method adopted for testing the sites, it was revealed that two sample sites were chosen in site 1 and 10 sample sites in site 2. Explain why this could have affected the results of the survey. **(3)**

(b) MediPop noticed that the poppy plants growing in the warehouse awaiting the new site had very different characteristics. Most of the poppies had orange petals but a few had very pale yellow petals. It was found that the pale yellow poppies did not produce berberine. The company carried out a dihybrid cross of a pure-bred orange poppy producing berberine (OOBB) with a pure-bred yellow poppy that did not produce berberine (oobb). All of the offspring were orange and produced berberine. An F2 cross was performed using the heterozygotes. The number of offspring of each phenotype are shown in the table.

Phenotype	**Number of offspring**
orange and berberine	139
yellow and berberine	5
orange and no berberine	3
yellow and no berberine	53

(i) The company hypothesised that the genes were located on the same chromosome. What is the scientific term for this? **(1)**

(ii) If the genes were on the same chromosome, a 3 : 1 ratio of orange and berberine to yellow and no berberine would have been expected. Draw a genetic figure to prove that this would be the case. **(2)**

(iii) Using the results shown in the table, state whether you agree with the company's hypothesis, and explain why. **(2)**

(iv) Suggest a possible reason for the presence of yellow/berberine and orange/no berberine plants in this cross. **(1)**

(Total for Question 20 = 14 marks)

21 Spider silk is a very versatile material that has many possible uses, including sutures in medicine. However, it is a difficult task to farm spiders. Some goats have been genetically engineered to produce spider silk in their milk.

(a) To isolate the spider silk gene the scientists first isolated the mRNA in the spider's silk-producing cells. What enzyme would they use to produce complementary DNA from the spider silk mRNA? **(1)**

(b) Outline the process of PCR that could be used to make further copies of the silk DNA. **(5)**

(c) Once sufficient copies of the spider silk gene are produced, it can be cut and inserted into a plasmid. Which two enzymes are necessary for this process? **(2)**

(d) The plasmid is then fused with a nucleus from an enucleated goat egg and, after fusion, the transgenic nucleus is injected into a second enucleated egg. This egg is then implanted into a third surrogate female goat. Suggest how the scientists could determine whether the genetic modification was successful before the goats are ready to produce milk. **(2)**

(e) Once the goat is genetically modified it will produce spider silk protein in its milk. Five embryos were genetically modified in this way and implanted into surrogate goats. The genetically engineered goats produced were milked, the spider silk protein was extracted from the milk, and the levels of spider silk protein were analysed.

(i) The volume of milk from which the protein was extracted was controlled for each goat. List three other controlled variables that would be necessary in this analysis experiment. **(3)**

(ii) The table shows the concentration of spider silk protein produced by each of the five goats in their milk.

Goat number	Spider silk protein/mg L^{-1}
1	11.7
2	10.3
3	16.1
4	4.2
5	10.6

Suggest a possible reason for the differences in concentration of silk protein produced, assuming all variables had been controlled. **(1)**

(f) A selection of genes in each of these goats was sequenced to try to determine the cause of the large differences in concentration of spider silk product. Goat number 4 had a deletion mutation in the gene coding for a transcription factor responsible for turning on expression of the engineered spider silk gene.

(i) How could this mutation affect the primary structure of the transcription factor? **(2)**

(ii) This mutation can alter the control of gene expression at the transcriptional level. Name one other level at which gene expression can be controlled and give an example of this type of control. **(2)**

(Total for Question 21 = 18 marks)

22 The domesticated chicken was selectively bred over 7000 years ago from red junglefowl (*Gallus gallus*). The domesticated chicken was bred for increased egg production and larger muscles for their meat.

(a) In the table, number the classification levels in the correct order and complete the last column. **(4)**

Order	Level	Name
	Species	
1	Kingdom	
	Phylum	Chordata
	Class	Aves
	Genus	
	Order	Galliformes
	Family	Phasianidea

(b) Use the information above, and your knowledge of natural selection, to explain how the domesticated chicken was selectively bred from red junglefowl. **(4)**

(Total for Question 22 = 8 marks)

OCR publishes official Sample Assessment Material on its website. This timed test has been written to help you practise what you have learned and may not be representative of a real exam paper.

Remember that the official OCR specification and associated assessment guidance materials are the only authoritative source of information and should always be referred to for definitive guidance.

A Paper 3

Time: 1 hour 30 minutes

Answer all the questions.

---

1 Phenylalanine hydroxylase (PAH) is an enzyme that catalyses the conversion of the amino acid phenylalanine into the amino acid tyrosine. PAH is only expressed in the liver. A section of the amino acid sequence of one of the subunits of PAH is shown in the figure.

121 FPRTIQELDR FANQILSYGA ELDADHPGFK DPVYRARRKQ FADIAYNYRH GQPIPRVEYM
181 EEEKKTWGTV FKTLKSLYKT HACYEYNHIF PLLEKYCGFH EDNIPQLEDV SQFLQTCTGF
241 RLRPVAGLLS SRDFLGGLAF RVFHCTQYIR HGSKPMYTPE PDICHELLGH VPLFSDRSFA
301 QFSQEIGLAS LGAPDEYIEK LATIYWFTVE FGLCKQGDSI KAYGAGLLSS FGELQYCLSE

(a) Part of the active site of PAH is highlighted in the amino acid sequence in the figure.

(i) Explain how these amino acids may end up being close together in the final polypeptide. **(4)**

(ii) Suggest what role these amino acids could have in the resultant protein. **(2)**

Phenylketonuria (PKU) is a metabolic disease characterised by little or no PAH enzyme activity. The figure shows the normal activity of PAH.

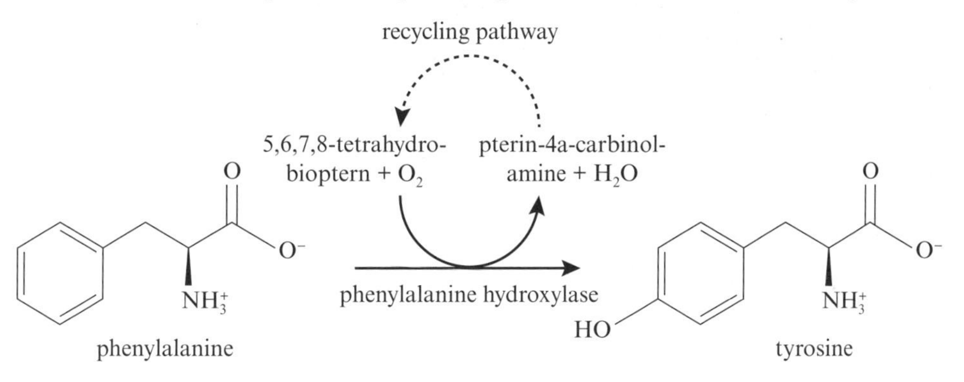

(b) What might be the consequences of reduced activity of PAH? **(2)**

(c) Samantha's mother and father are both carriers of the PKU mutation. They have three daughters besides Samantha, none of whom have the disease although two are carriers of the PKU mutation. One of Samantha's sisters, Celine, has two children fathered by a carrier of the PKU mutation. Two of these children have the disease.

(i) State what pattern of inheritance PKU shows in this family. **(1)**

(ii) If Celine were to have a third child with the same father, what is the probability this child will have the disease? **(1)**

To determine the activity of Samantha's mutated PAH, the mutated version of the enzyme was produced using *in-vitro* gene expression and isolated. Entrapment was used to immobilise the enzyme and an assay carried out to determine the levels of phenylalanine and tyrosine present in a solution.

(d) Describe how the activity of the mutated PAH enzyme could be evaluated experimentally using entrapment. **(5)**

(Total for Question 1 = 15 marks)

2 Saxitoxin is a neurotoxin that is produced by certain species of marine dinoflagellates and cyanobacteria. If ingested, this toxin can be fatal to humans.

A group of researchers wished to investigate the levels of saxitoxin present in bivalve molluscs that are known to feed on algal blooms of the dinoflagellate genus *Alexandrium*, the main genus that produces this toxin. They wanted to determine whether seasonal changes affected the levels of saxitoxin in the molluscs.

- A sample of 30 bivalve molluscs was taken from a sandy shore in summer.
- Saxitoxin levels were determined using a standardised laboratory technique and recorded for all of the organisms collected.
- The procedure was repeated in winter.
- All the results were recorded in the table.

Season	Saxitoxin levels/µg per 100 g tissue						Range	Mean	SD
summer	78.26	78.02	79.23	78.38	77.97	75.01			
	76.34	78.48	65.08	67.73	73.73	78.78			
	67.45	75.67	79.93	70.97	74.87	72.29			
	72.32	76.98	79.34	69.70	69.85	65.03			
	73.64	75.65	74.13	67.65	76.18	65.24	14.90	73.80	4.69
winter	12.71	15.67	17.83	17.51	15.21	10.24			
	10.06	13.68	10.10	15.75	16.02	12.60			
	18.39	10.18	18.07	14.99	19.21	14.04			
	17.94	12.74	17.76	13.79	13.80	12.73			
	18.25	17.48	13.80	14.93	12.61	14.74	9.15	14.76	2.72

(a) The *t*-test can be carried out to determine the significance of the differences between the samples collected in different seasons.

(i) Calculate the *t* value for the data using the following formula:

$$t = \frac{|\bar{x}_1 - \bar{x}_2|}{\sqrt{\left(\frac{s_1^2}{n_1} + \frac{s_2^2}{n_2}\right)}}$$

where, $|\bar{x}_1 - \bar{x}_2|$ is the difference in mean values of sample 1 and sample 2, s_1^2 and s_2^2 are the squares of the standard deviations of the samples, and n_1 and n_2 are the sample sizes. Give your answer to decimal places. **(2)**

(ii) State the null hypothesis for these data. **(1)**

(iii) The critical values at 58 degrees of freedom are shown in the table.

degrees of freedom	$p = 0.10$	$p = 0.05$	$p = 0.01$	$p = 0.001$
58	1.67	2.00	2.66	3.47

Using the table of critical values, explain the significance of the data presented above. **(2)**

(b) Suggest a limitation of the procedure used to gather data in this experiment and how you could improve this. **(2)**

(c) Discuss the validity of this experiment. **(3)**

(d) Saxitoxin is a major cause of paralytic shellfish poisoning (PSP) as it acts on the voltage-gated sodium channels of neurones. Explain how this could cause paralysis. **(5)**

(Total for Question 2 = 15 marks)

3 Naked mole-rats (*Heterocephalus glaber*) live in burrows in arid deserts and are mammals that show eusocial behaviour, meaning that there is division of labour creating separate behavioural groups within the population, called castes.

(a) The figure shows a naked mole-rat.

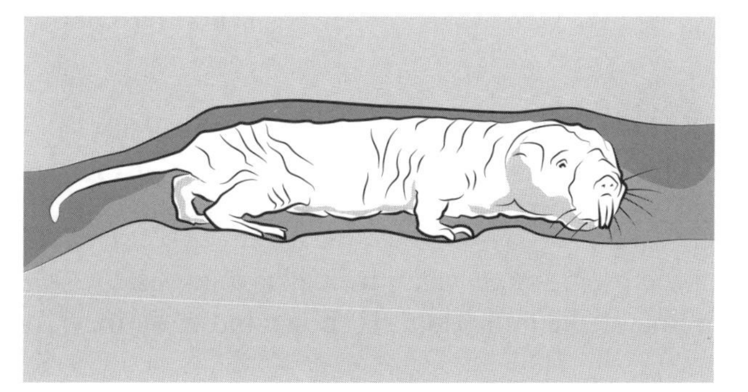

Using the physical characteristics visible in the figure, suggest two adaptations of the naked mole-rat and how each could confer an advantage in the organism's environment. **(2)**

(b) In addition to behavioural and physical adaptations, naked mole-rats display physiological adaptations, including resistance to pain. The cause of pain resistance is a lack of a neurotransmitter called substance P (SP), which acts in a similar way to the neurotransmitter acetylcholine. Explain how SP is normally involved in the transmission of a pain signal from the sensory receptors in the skin to the brain? **(2)**

(c) The ribosomes of naked mole-rats have been shown to produce mostly error-free polypeptides. What is the name of the process by which ribosomes produce polypeptides? **(1)**

(d) Naked mole-rats also appear to be unusually resistant to cancer. A possible cause of this is postulated to be due to a protein called hyaluronan, which is mainly responsible for the increased elasticity of the mole-rat skin.

(i) The table shows the levels of hyaluronan (HA) in naked mole-rats and three other organisms compared to a positive control. Plot the most suitable graph of the data given in the table. **(4)**

	Relative hyaluronan (% of control)	**Standard deviation**
control	100	±0
naked mole-rat	100	±2
mouse	62	±5
guinea pig	58	±6
human	10	±4

(ii) Use the data in the table to describe the differences in the levels of hyaluronan in the organisms studied. **(2)**

(iii) If higher levels of HA are indeed responsible for increased cancer resistance, what could be said about the resistance of the other three organisms? **(1)**

(Total for Question 3 = 12 marks)

4 Due to the high number of applications for sports scholarships, a university made the decision to introduce admission fitness tests to ensure intake of the best possible applicants in order to boost the reputation of the university.

(a) One of the tests included in the programme was spirometric analysis. A representative trace from one of the individuals is shown in the figure.

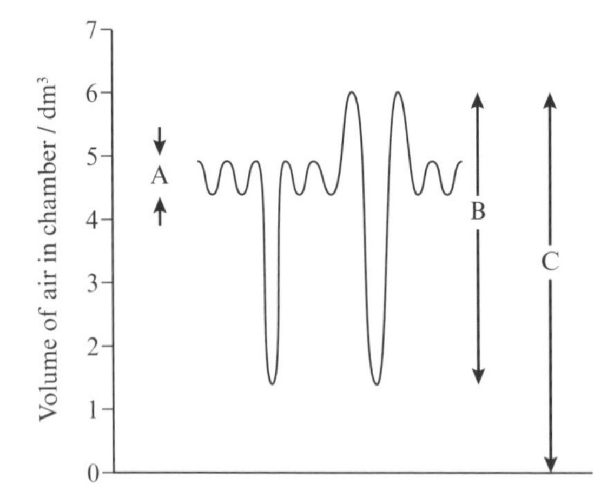

(i) Labels A, B and C indicate measurements that can be determined using a spirometer trace. Give the names of these measurements. **(2)**

(ii) Athlete X had a resting breathing rate of 14 breaths per minute and athlete Y of 18 breaths per minute. Suggest two reasons why the breathing rates of the two athletes were different. **(2)**

(b) During spirometry the volume of gas in the chamber decreases during the experiment. For athlete X the volume in the chamber decreased from 4.8 dm^3 to 3.6 dm^3 in 30.6 seconds. For athlete Y the volume in the chamber decreased from 4.8 dm^3 to 3.7 dm^3 in 30.1 seconds.

(i) Why does the volume of gas in the chamber decrease? **(1)**

(ii) Calculate the rate of oxygen consumption per second for athletes X and Y to three decimal places. **(2)**

(iii) What further information would be useful to determine the relative fitness levels of these athletes? **(1)**

(iv) Explain why a high rate of oxygen consumption could indicate high levels of fitness? **(2)**

(c) The admissions tests for the university also involve a 1 mile run and a lactate test where athletes are asked to cycle at increasing speeds until the level of lactate detectable in their blood reaches more than 4 mmol L^{-1}. Another athlete applying for the scholarship indicated on her medical form that she suffers from a mild form of pyruvate dehydrogenase complex deficiency (PDCD). The pyruvate dehydrogenase complex consists of three enzymes that function to convert pyruvate into acetyl-CoA. Explain why and how a deficiency in this complex could affect this athlete's performance in the university's admission tests. **(5)**

(Total for Question 5 = 15 marks)

5 In October 2013, badger culling was piloted in two areas of the UK to tackle the rising cases of bovine tuberculosis (bTB) amongst beef and dairy sectors. There was evidence to suggest that badgers were carrying bTB between cattle and contributing to the spread of the disease. However, other animals, such as cats and deer, have also been shown to transmit the disease and badgers are not tested for bTB before culling.

(a) To what kingdom does the pathogen that causes TB belong? **(1)**

(b) To determine the prevalence of bTB in a badger population in the vicinity of a farm with contaminated cattle, some researchers baited badger traps with peanuts and set them up overnight. Each trap could hold only one badger and blood samples were taken from the badgers in the morning before release. What methods should the researchers have employed to ensure that they obtained a representative sample of the badger population and minimal disruption was made to the population? **(3)**

(c) The blood samples collected were processed and the numbers of badgers with bTB in this area were noted along with those in the vicinity of a farm without bTB contamination. In an average population of badgers in the UK the prevalence of TB is estimated to be around 15%. The table of results is shown below.

Sampling area	Number of badgers with TB	Number of badgers without TB
near bTB farm	10	40
near non-bTB farm	26	174

The chi-squared formula is:

$$\sum \frac{(O - E)^2}{E}$$, where O = observed values and E = expected values.

Critical values for the chi-squared test can be found on page 141 of the Revision Guide.

(i) Use the information above and your knowledge of the chi-squared test to determine whether the prevalence of bTB near bTB farms is significantly different from the expected UK average. Show your working, including the expected values for TB and non-TB badgers and your test value. **(3)**

(ii) Why might it be inappropriate to make a direct comparison between the prevalence of TB in the badgers near bTB farms and near non-bTB farms considering the data above? **(1)**

(d) TB is an airborne pathogen. Describe what defences animals have against such microorganisms. **(5)**

(Total for Question 6 = 13 marks)

Answers

The answers provided here are examples and provide key points. In some cases, other answers may also be possible.
In your exam you should aim to write in full sentences.

1. Using a light microscope

1 Multiply the magnification of the objective lens **(1)** by the magnification of the eyepiece lens. **(1)**

2 (a) false **(1)** (b) true **(1)** (c) true **(1)** (d) true **(1)**

3 (a) three from:
cell wall **(1)** nucleus **(1)** chloroplast **(1)** cytoplasm **(1)**
(b) The maximum magnification of the light microscope is ×1500, and the resolution is 200 nm. **(1)**
This magnification and resolution are too low to observe most organelles of the cell. **(1)**

2. Using other microscopes

1 **B (1)**

2 Magnification is the number of times larger the image is than the specimen. **(1)** Resolution is the ability to distinguish between two points on an image. **(1)** The higher the resolution, the sharper the image. **(1)**

3 An electron microscope has much better resolution. However, specimens cannot be alive and there is so much harsh treatment in the preparation of materials it is not always clear what is real and what is artefact.
A light microscope has limited resolution so magnification above about 1000× is pointless. Many organelles are too small to see at this power. However, living material can be viewed and the problem of artefacts is much less severe.

3. Preparing microscope slides

1 (a) 17 **(1)** (b) 1.7 mm **(1)**

2 **C (1)**

3 Cut a very thin slice of the specimen. **(1)**
Stain the specimen. **(1)**
Place a coverslip over the stained specimen. **(1)**

4. Calculating magnification

1 **B (1)**

2 magnification $= \frac{15{,}000\,\mu m}{2\,\mu m}$ **(1)** = ×7,500 **(1)**

3 (a) nuclear pore **(1)**
(b) object size $= \frac{\text{image size}}{\text{magnification}}$ **(1)**
$= \frac{4000\,\mu m}{25000}$ **(1)** $= 0.16\,\mu m$ **(1)**

5. Eukaryotic and prokaryotic cells

1

	Present in animal cells	Present in plant cells
nucleus	✓	✓
nucleolus	✓	✓
rough endoplasmic reticulum	✓	✓
smooth endoplasmic reticulum	✓	✓
ribosomes	✓	✓
mitochondria	✓	✓
chloroplasts	✗	✓
Golgi apparatus	✓	✓
centrioles	✓	✗
cellulose cell wall	✗	✓
plasma membrane	✓	✓
large central vacuole	✗	✓
cytoskeleton	✓	✓

1 mark for each 4 correct responses

2 (a) true **(1)** (b) false **(1)** (c) false **(1)** (d) true **(1)**

3 The rough endoplasmic reticulum synthesises proteins. **(1)**
The smooth endoplasmic reticulum synthesises lipids. **(1)**
The Golgi apparatus modifies proteins for exocytosis. **(1)**

6. The secretion of proteins

1 Proteins are made by the ribosomes. **(1)**
Proteins are made in the rough endoplasmic reticulum. **(1)**
The proteins are packaged into vesicles. **(1)**
These travel to the Golgi apparatus. **(1)**
This organelle then modifies the proteins and packages them into vesicles. **(1)**
Finally, they are carried to the plasma membrane. **(1)**

2 **C (1)**

3 (a) The beta cells of the pancreas produce insulin, a protein. **(1)**
The rough endoplasmic reticulum (ER) contains ribosomes. **(1)**
It is the ribosomes that synthesise proteins. **(1)**
(b) Proteins are secreted from the beta cells by exocytosis, **(1)** which requires ATP / is an active process. **(1)**

7. The cytoskeleton

1 **B (1)**

2 (a) Flagella: role is movement of the cell. **(1)**
(b) Cilia: role is movement of the cell / movement of substances, e.g. mucus. **(1)**
(c) Microtubules: role is to maintain the shape of the cell. **(1)**
(d) Microtubule motors: role is movement of vesicles along microtubules. **(1)**

3 Microtubule motors move vesicles along microtubules inside the cell. **(1)**
Flagella / cilia are made from microtubules on the surface of the cell which help the cell to move. **(1)**
The lengthening and shortening of microtubules inside the cell helps cells to move. **(1)**

8. The properties of water

1 oxygen **(1)**, hydrogen **(1)**, polar **(1)**, negative **(1)**, positive charge **(1)**

2 **A (1)**

3 Water is a solvent and can transport many substances in the blood. **(1)** Most reactions in the body are carried out in the cytoplasm of cells, which is mostly water. **(1)** Water is used in metabolic reactions, to make and break bonds. **(1)**

4 Ice is dense and floats on the sea, allowing animals to live below the ice.
Water is locked up as ice and is unavailable for use by plants. **(1)**

9. The biochemistry of life

1 **A (1)**

2 **C (1)**

3 Water is used in hydrolysis to break bonds. **(1)**
The hydrogen atom / H of the water binds to the oxygen in the centre of the bond, and the hydroxyl group / OH binds to a carbon on the other side of the bond. **(1)**
Water is also used in condensation reactions to make bonds. **(1)**
A hydrogen atom / H is removed from a hydroxyl group / OH on one side of the bond and a hydroxyl group / OH is removed from the other side of the bond, and together they form a molecule of water. **(1)**

10. Glucose

1 D (1)

2 galactose (1) disaccharide (1)
The lactase enzyme breaks the glycosidic bond between these two monoscaccharides (1)

3 People who are lactose intolerant do not have the enzyme lactase, (1) which is needed to digest lactose. (1) The bacteria in the yoghurt contain an enzyme that digests the lactose in the yoghurt. (1) The lactose is digested into glucose and galactose. (1)

11. Starch, glycogen and cellulose

1 C (1)

2 Glucose molecules are joined together by glycosidic bonds. (1)
Starch and glycogen molecules are highly branched. (1)
Both starch and glycogen are easily hydrolysed by enzymes. (1)

3 Starch is made from alpha glucose. (1)
Cellulose is made from beta glucose. (1)
Starch is made from amylopectin and amylose and is coiled and highly branched. (1)
Cellulose is made of long straight chains held together by hydrogen bonds. (1)
Starch is an energy storage molecule. (1)
Cellulose gives strength to plant cell walls. (1)

12. Triglycerides and phospholipids

1

	Triglyceride	Phospholipid	
number of ester bonds	3	2	(1)
number of fatty acid chains	3	2	(1)
presence of a glycerol molecule?	✓	✓	(1)
presence of a phosphate group?	✗	✓	(1)

2 B (1)

3

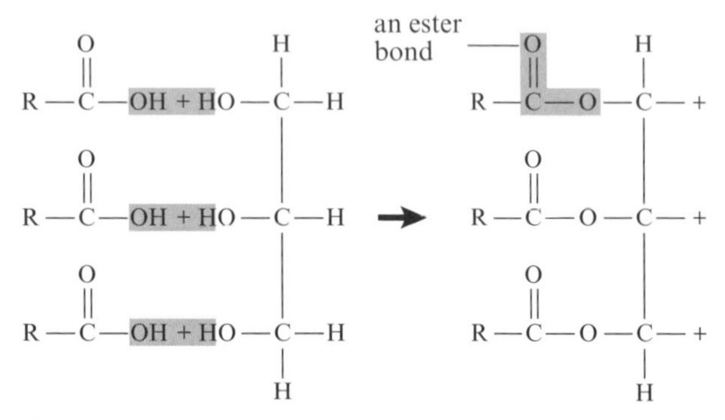

glycerol molecule correctly drawn (1)
at least one fatty acid molecule correctly drawn (1)
at least one ester bond correctly shown (1)

13. Use of lipids in living organisms

1 A (1)

2 Phospholipids create a barrier between the inside of the cell and the environment. (1)
Phospholipids prevent large, polar molecules passing through the plasma membrane. (1)
Cholesterol stabilises plasma membranes and prevents them from freezing. (1)
Glycolipids aid in cell signalling and cell recognition on the cell surface. (1)

3 Aquatic mammals use triglycerides for buoyancy on the water. (1)
They also use triglycerides to make a thick layer of blubber as insulation from the cold water. (1)
Aquatic animals need a lot of energy for swimming and use triglycerides as an energy store. (1)

14. Amino acids

1 C (1)

2 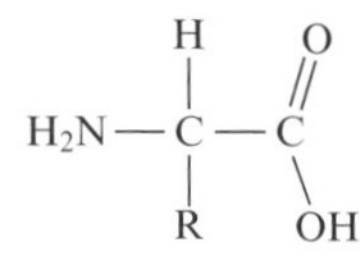 (1) for H_2N– and (1) for –COOH

3 Most amino acids can be made by the human body; (1) however, essential amino acids cannot be made in the body and have to be acquired through diet. (1)

4 (a) A peptide bond forms between the carboxyl group / COOH of one amino acid and the amino group / NH_2 of another amino acid. (1)
A hydroxyl group / OH and a hydrogen atom / H break off from their amino acids to form a molecule of water. (1)

(b) When a molecule of water is added to the peptide bond (1) a hydroxyl group / OH is added to the carboxyl group / COOH of one amino acid and a hydrogen atom / H is added to the amino group of the other amino acid / NH_2, breaking the bond. (1)

15. Protein

1 (a) Compare: same types of chemical bonds within the amino acids and between amino acids (1)
Contrast: secondary structure formed by hydrogen bonding into alpha helix and beta sheet (1)

(b) Compare: both have a precise shape held in place by bonds between amino acids (disulfide links, H-bonds, ionic bonds, hydrophobic/hydrophilic interactions) (1)
Contrast: quaternary involves more than one polypeptide chain (1)

2 B (1)

3 Hydrogen bonds can form between the hydrogen atoms and the oxygen atoms of adjacent amino acids. (1)
Disulfide bridges are formed between two atoms of sulfur. (1)
Only the amino acid cysteine contains sulfur, so a disulfide bridge can only form when two cysteine amino acids are in close proximity. (1)
An ionic bond is formed between a negatively charged amino acid and a positively charged amino acid. (1)
These bonds can only form in the tertiary or quaternary structure (1) when a negatively charged amino acid and a positively charge amino acid are in close proximity. (1)

16. Fibrous proteins

1 Collagen in tendons / cartilage (1)
provides great mechanical strength. (1)
Keratin in hooves / horns / fingernails (1)
is hard and strong. (1)
Elastin in skin / blood vessels (1)
is elastic / allows structures to stretch. (1)

2 A globular protein is roughly spherical and may have a quaternary structure. (1)
A fibrous protein is long and thin and may also have a quaternary structure. (1)
Globular proteins may contain prosthetic groups. (1)

3 Collagen is used in the tendons, to attach the muscles to the bones, and in the ligaments, to attach bones to other bones. (2)

17. Globular proteins

1 A prosthetic group is a part of a protein that is not made of amino acids. (1)

2 A (1)

3 (a) contains a haem group that binds to oxygen (1)

(b) Haemoglobin is made of four polypeptide chains.
Haemoglobin has a quaternary structure. (1)
Each polypeptide chain is folded into a globular subunit / tertiary structure. (1)
Each polypeptide chain / subunit contains a haem group containing an Fe^{2+} ion. (1)
The haem group is a prosthetic group. (1)

(c) A decrease in iron would lead to a decrease in haemoglobin, which is needed to carry oxygen in the blood. (1)

18. Benedict's test

1 **B (1)**
2 Benedict's reagent contains copper(II) ions. **(1)**
Reducing sugars, such as glucose, reduce the copper(II) ions to copper(I) ions. **(1)** This forms a red precipitate. **(1)**
3 Orange juice contains fructose, which is a reducing sugar. **(1)**
The Benedict's test for reducing sugars should be used. **(1)**
The copper(II) ions in Benedict's reagent will be reduced by the fructose, **(1)** producing a red precipitate. **(1)**

19. Tests for proteins, starch and lipids

1

Test	Positive	Negative
Biuret's test	mauve **(1)**	blue
Iodine test	blue-black **(1)**	orange **(1)**
Emulsion test	white emulsion	no emulsion **(1)**

2 **D (1)**
3 (a) Positive result for biuret's test, as milk contains proteins. **(1)**
Negative result for iodine test, as milk does not contain starch. **(1)** Positive result for emulsion test, as milk contains fat. **(1)**
(b) Positive result for iodine test, as potato contains starch. **(1)**
Negative result for emulsion test, as potato does not contain fat. **(1)**

20. Practical techniques – colorimetry

1 **C (1)**
2 A D G F E C B **(4)**
3 Samples of known concentrations of iron could be used to measure the absorbance at each concentration using a colorimeter **(1)** and a calibration curve drawn. **(1)** Absorbance could be measured in the samples taken from the river. **(1)** The calibration curve could be used to work out the concentration of iron in the river samples. **(1)**

21. Practical techniques – chromatography

1 **B (1)**
2 (a) R_f values are:
B = 0.4 **(1)** C = 0.33 **(1)** D = 0.4 **(1)**
(b) B and D could be the same compound as their R_f values are the same. **(1)**
3 Dot the sample of ink on the chromatography paper. **(1)**
Dry the spot and re-spot, to increase concentration. **(1)**
Place the paper in a jar with a solvent that the components will dissolve in, just touching the bottom of the paper. **(1)**
The solvent will move up the paper, separating the components of the ink. **(1)**
Components with the lowest affinity for the paper will be at the top of the paper, and components with the highest affinity for the paper will be at the bottom of the paper. **(1)**

22. Exam skills

1 (a) A mitochondrion **(1)**
B cellulose cell wall **(1)**
C chloroplast **(1)**
D smooth endoplasmic reticulum / smooth ER **(1)**
(b) The site of aerobic respiration. **(1)**
(c) The concentrated salt solution has a lower water potential than the cell. **(1)**
Water moves out of the cell by osmosis. **(1)**
Water moves from higher water potential inside the cell, to lower water potential outside the cell. **(1)**
The plant cell will eventually become plasmolysed. **(1)**

23. Nucleotides

1 (a) X = phosphate group **(1)** Y = deoxyribose sugar **(1)**
Z = thymine / T **(1)**
(b) **A (1)**
2 (a) messenger RNA / mRNA **(1)**
(b) The hydrogen atom is removed from the hydroxyl group on carbon 3 of the ribose sugar of one nucleotide. **(1)**
The hydroxyl group is removed from the phosphate group of the other nucleotide. **(1)**
A covalent bond is formed between carbon 3 of the ribose and the phosphate group. **(1)**
(c) hydrolysis **(1)**

24. ADP and ATP

1 **D (1)**
2 (a) ATP and RNA nucleotides both contain a nitrogenous base **(1)** and a ribose sugar. **(1)**
ATP and RNA nucleotides contain different numbers of phosphate groups: ATP has three phosphate groups and RNA nucleotides contain only one phosphate group. **(1)**
(b) two from: contraction of muscles **(1)**
active transport across plasma membranes **(1)**
glycolysis **(1)**
(c) The phosphoanhydride bond between the second and third phosphate is hydrolysed. **(1)**
This forms ADP and an inorganic phosphate. **(1)**
30.5 kJ mol^{-1} of energy is released. **(1)**

25. The structure of DNA

1 **C (1)**
2 (a) ATGGAAGTAAATCCC **(1)**
(b) A / adenine always pairs with T / thymine **(1)**
C / cytosine always pairs with G / guanine **(1)**
There are two hydrogen bonds between adenine and thymine, and three hydrogen bonds between cytosine and guanine. **(1)**
3 The cell surface membrane and nuclear envelope are physically broken down. **(1)**
The mixture is filtered. **(1)**
The DNA is precipitated from solution using ethanol. **(1)**

26. Semi-conservative DNA replication

1 **B (1)**
2 (a) After replication, each DNA molecule contains one strand of the original DNA **(1)** and one strand of new DNA. **(1)**
(b) DNA helicase breaks the hydrogen bonds between the complementary base pairs. **(1)**
DNA polymerase adds new DNA nucleotides to each strand. **(1)**
The DNA nucleotides added to the new strand of DNA are complementary to the nucleotides on the original DNA strand. **(1)**
(c) Mistakes in the DNA can lead to mutations. **(1)**
Mutations can lead to the development of proteins with different shapes. **(1)** These proteins may be ineffective and could lead to tumours. **(1)**

27. The genetic code

1 The genetic code is degenerate, which means that several different triplets can code for the same amino acid. **(1)**
The genetic code is universal, which means that all organisms have the same triplets for the same amino acids. **(1)**
The genetic code is non-overlapping, and so is read from a fixed point in groups of three bases. **(1)**
2 Met Gly Cys Tyr His **(1)**
3 causes a frameshift in the genetic code **(1)**
causes a change in the amino acid sequence **(1)**
could result in a non-functional or truncated protein **(1)**

28. Transcription and translation of genes

1 **A (1)**
2 (a) RNA nucleotides **(1)** are added to the template strand of DNA in a complementary way **(1)** by RNA polymerase. **(1)**
(b) five from:
Each tRNA molecule carries one amino acid. **(1)**
Each tRNA molecule has an anti-codon. **(1)**
The tRNA travels to the ribosome. **(1)**

The anti-codon pairs up in a complementary way with the codon on the mRNA. **(1)**
The amino acid from the tRNA is added to the polypeptide. **(1)**
The amino acids are joined by a peptide bond. **(1)**

29. The role of enzymes

1 C **(1)**

2 (a) (i) enzyme found inside cells **(1)** (ii) catalase **(1)**
(b) Trypsin is packaged into vesicles that travel to the cell surface membrane. **(1)**
Vesicles fuse with the membrane and secrete trypsin outside of the cell. **(1)**
Secretion into the small intestine is by exocytosis. **(1)**

3 The amylase would be denatured as the hydrogen bonds holding the tertiary structure together would break. The active site of the amylase would no longer function. **(2)**

4 Enzymes provide a lower energy pathway to convert substrates into products.
They do this by binding with the substrate, and providing an environment that makes the reaction more likely to occur.
This allows reactions to occur without the need to raise the temperature of the cell. **(3)**

30. The mechanism of enzyme action

1 Each active site is specific to its substrate. **(1)**
The shape of the active site of each enzyme is complementary to its substrate. **(1)**
Only the correct substrate will fit into the active site. **(1)**

2 C **(1)**

3 The sucrase and sucrose form an enzyme–substrate complex. **(1)**
The sucrase breaks down the sucrose to form the products / glucose and fructose. **(1)**
The sucrase and the products are now called the enzyme–product complex. **(1)**

31. Factors that affect enzyme action

1 B **(1)**

2 (a) The temperature at which the enzyme has the highest rate of reaction. **(1)**
(b) Every time the temperature increases by 10 °C, the activity of the enzyme roughly doubles. **(1)**
As the temperature increases, the kinetic energy of the enzyme and substrate increases. **(1)**
The enzyme and substrate will collide more often, resulting in more enzyme–substrate complexes. **(1)**

32. Factors that affect enzyme action – practical investigations

1 (a) 40 °C **(1)**
(b) $8\,cm^3\,s^{-1}$ **(1)**
(c) $\frac{(8-2)}{2} \times 100\%$ **(1)**
$= 300\%$ **(1)**
(d) After 40 °C, the rate of oxygen production decreases. **(1)**
This is due to denaturation of catalase / breaking of hydrogen bonds in the active site of catalase. **(1)**
Catalase is no longer able to bind to hydrogen peroxide / substrate / form an enzyme–substrate complex. **(1)**
(e) A **(1)**

33. Cofactors, coenzymes and prosthetic groups

1 A **(1)**

2 (a) Cofactors are non-protein inorganic substances.
Cofactors are needed for the enzyme reaction to occur at an appropriate rate. **(1)**
(b) A co-substrate binds to the enzyme, activating the enzyme. **(1)**
The co-substrate allows the substrate to bind to the enzyme's active site. **(1)**

3 (a) two from:
haem group in haemoglobin, binds oxygen **(1)**
magnesium ion in chlorophyll, essential for photosynthesis **(1)**
zinc ion in carbonic anhydrase, needed to convert CO_2 and H_2O into H_2CO_3 **(1)**
(b) Minerals are needed as cofactors. **(1)**
Vitamins are needed as coenzymes. **(1)**

34. Inhibitors

1 (a) non-competitive inhibitor **(1)**
As the substrate concentration increases, the rate of reaction with the inhibitor remains lower than the rate of reaction without, the inhibitor, **(1)**
indicating that the inhibitor does not compete with the substrate for the active site. **(1)**
(b) The inhibitor binds to an allosteric site on the enzyme. **(1)**
This binding changes the shape of the active site **(1)**
preventing the substrate from binding there. **(1)**

2 D **(1)**

35. Exam skills

1 (a)

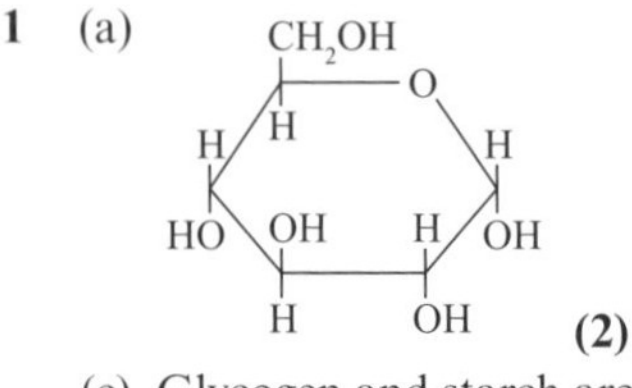

(2)

(b) (Alpha) 1,6 glycosidic bond **(1)**
(c) Glycogen and starch are both made from alpha glucose molecules.
Glycogen and starch are both coiled and highly branched.
Glycogen contains only one polysaccharide and starch contains amylopectin and amylose. Glycogen has many short branches whereas starch has fewer long branches. **(4)**
(d) Add Benedict's reagent to the sample. Heat the sample. After a few minutes, the sample will change colour to green, yellow, orange or brick-red if a reducing sugar / glucose is present. The sample will remain blue if there is no reducing sugar / glucose present. **(3)**
(e) There is a small amount of glucose present in the sample. **(1)**

36. The fluid mosaic model

1 D **(1)**

2 (a) a membrane protein that is permanently attached to the plasma membrane **(1)**
(b) channel protein: allows substances to diffuse across the membrane **(1)**
carrier protein: actively transports substances across the plasma membrane **(1)**

3 Hormones bind to receptor proteins **(1)** in a complementary way. **(1)** This causes a series of reactions inside the cell. **(1)**

37. Factors that affect membrane structure

1 Cholesterol partially immobilises nearby phospholipids in the plasma membranes. **(1)**
This decreases the permeability and maintains the mechanical stability of the plasma membrane. **(1)**

2 B **(1)**

3 (a) As the temperature increases from 0 °C to 60 °C, the percentage of cells containing methylene blue increases. **(1)**
At 0 °C, 0% of cells contain methylene blue; at 60 °C, 54% of cells contain methylene blue. **(1)**
(b) As the temperature increases above 40 °C, carrier proteins would start to denature.
If carrier proteins were involved then uptake would be expected to reduce at temperatures above 40 °C. **(1)**
Instead, the data show that uptake continues as temperature continues to rise, so carrier proteins are not involved. **(1)**

38. Movement across the membrane

1 D **(1)**

2 (a) Glucose can be transported from an area of high concentration outside of the cell to an area of low concentration inside the cell **(1)** by facilitated diffusion. **(1)** Glucose can also be transported against the concentration gradient by active transport. **(1)** The intermediate energy source for active transport is ATP. **(1)**
(b) Phospholipids act as a barrier. **(1)**
Glucose is too large to pass between the phospholipids of the plasma membrane. **(1)**
3 (a) exocytosis **(1)**
(b) Phagocytosis takes substances into the cell and exocytosis takes substances out of the cell. **(1)**
Phagocytosis is the engulfing of pathogens or foreign substances in the body. **(1)**

39. Osmosis

1 (a) Water moves **(1)** from an area of high water potential to an area of low water potential **(1)** across a partially permeable membrane. **(1)**
(b) Water moves out of the plant cells **(1)** causing them to become plasmolysed. **(1)**
2 (a) Water potential is a measure of how free the water molecules are to move from one area to another. **(1)**
It is a type of pressure and is measured in kilopascals (kPa). **(1)**
(b) so that water moves into the plant cells **(1)** and the cells become turgid and maintain the cells' shape and stability **(1)**
3 **A (1)**

40. Factors affecting diffusion

1 **C (1)**
2 (a) Potato chips gained mass as water moved into the potato chips if the water potential of the potato chips was lower than the water potential of the sugar solution they were placed into. **(1)**
Potato chips lost mass as water moved out of the potato chips if the water potential of the potato chips was higher than the water potential of the sugar solution they were placed into. **(1)**
(b) The change in mass of the potato chips is plotted on the *y*-axis of a line graph. **(1)**
The sugar concentration is plotted on the *x*-axis. **(1)**
The sugar concentration of the potato chips is where the plotted line crosses the *x*-axis. **(1)**
(c) (i) three from:
mass of the potato chips
shape / surface area of the potato chips **(1)**
variety / age of potato **(1)**
time potato chips are in sugar solution **(1)**
volume of sugar solution **(1)**
(ii) two from:
not using the same balance each time to weigh the potato chips **(1)**
judging the meniscus while using a measuring cylinder to measure the volume of the sugar solutions **(1)**
not all of the excess moisture removed from the potato chips before weighing **(1)**

41. The cell cycle

1 **A (1)**
2 (a) Hayflick limit **(1)**
(b) stem cells **(1)** tumour cells **(1)**
(c) senescence, cells no longer divide **(1)**
programmed cell death / apoptosis **(1)**
3 (a) The cell cycle is regulated by a regulatory protein, called p53, which acts at the G1/S checkpoint and the G2/M checkpoint. p53 checks the DNA for damage / mismatched bases / mutations. **(1)** If there is any DNA damage, the cell is not allowed to progress through the cell cycle until the damage is repaired. **(1)**
(b) Mutations in the DNA could lead to non-functioning proteins. **(1)** A non-functioning p53 protein would allow cells with damaged DNA through the cell cycle and lead to tumours developing. **(1)**

42. Mitosis

1 **B (1)**
2 (a) contains two sets of chromosomes **(1)**
(b) Nuclear envelope: breaks down **(1)**
Centrioles: divide and move to the poles of the cell, start to make the spindle fibres **(1)**
Chromosomes: supercoil and become more visible **(1)**
3 (a) asexual reproduction in some plants, e.g. strawberries, daffodils **(1)**
growth of roots and shoots of plants **(1)**
repair of damaged parts of a plant **(1)**
(b) root tip or shoot tip **(1)**

43. Meiosis

1 **C (1)**
2 prophase I **(1)** anaphase I **(1)**
metaphase I **(1)** telophase I **(1)**
3 (a) metaphase I **(1)**
(b) Six single chromosomes shown, three on each side of the cell **(1)** moving away from the equator towards the poles of the cell. **(1)**
(c) Homologous chromosomes line up together on the equator of the cell. **(1)**
Chromosome arms cross over at each chiasma. **(1)**
DNA from that section of the chromosome is exchanged for DNA on the other chromosome. **(1)**

44. Specialised cells

1 **B (1)**
2 Palisade cells are long and thin so that they can be closely packed together. **(1)** Palisade cells are packed with chloroplasts to absorb sunlight. **(1)** Guard cells work in pairs and have a thick cellulose cell wall on one side **(1)** to be able to open and close the stomata between them. **(1)**
3 (a) Sperm cells have a long undulipodium to swim to the ovum. **(1)**
The head of the sperm cell contains an acrosome, which is packed with enzymes to digest the wall of the ovum and allow the head of the sperm to enter the ovum. **(1)**
The sperm is packed with mitochondria to provide energy for the undulipodium to swim to the ovum. **(1)**
(b) To make sure that the chromosome number of the zygote does not increase. **(1)**

45. Specialised tissues

1 **C (1)**
2 (a) Ciliated epithelium contain cilia that waft mucus out of the lungs. **(1)**
Muscle tissue is made of many muscle cells that contract together in the same direction. **(1)**
(b) Squamous and ciliated epithelium both line inside surfaces.
Squamous epithelium is smooth **(1)**
and forms the inside surface of blood vessels and alveoli. **(1)**
Ciliated epithelium contains cilia and goblet cells **(1)**
and lines the inside of the trachea and bronchi. **(1)**
3 Water moves through the xylem **(1)**
from the roots up to the leaves. **(1)**
Sucrose moves through the phloem **(1)**
up and down the plant. **(1)**

46. Stem cells

1 **A (1)**
2 (a) A ball of cells formed in the early embryo **(1)**
made of a central mass of pluripotent cells, surrounded by a sphere of stem cells that will develop into the placenta and umbilical cord. **(1)**
(b) Pluripotent cells from the inner cell mass of the blastocyst are removed. **(1)**

Cells are treated with chemicals to stimulate them into differentiating into different cell types. **(1)**

3 (a) Plant stem cells are found in the meristem in the root **(1)** and shoot tips. **(1)**
Stem cells are also found in the cambium between the xylem and the phloem in the vascular bundle. **(1)**
(b) Stem cells in the cambium in the vascular bundle **(1)** differentiate to form xylem and phloem tissue. **(1)**

47. Uses of stem cells

1 **D (1)**
2 (a) Pluripotent / can differentiate into any type of body cell. **(1)**
Embryonic stem cells are more plentiful **(1)**
and are easier to harvest than other types of stem cell. **(1)**
(b) unethical to use embryonic stem cells as they could develop into a fetus if implanted into a uterus **(1)**
(c) adult stem cells that have been chemically induced to undifferentiate back into pluripotent stem cells **(1)**
3 (a) Stem cells are chemically induced to differentiate into one of the organ's tissues **(1)**
and grown over a scaffold in the shape of the organ. **(1)**
(b) Using stem cells to grow new organs allows people to receive an organ made from their own cells, reducing the chances of organ rejection. **(1)**

48. Exam skills

1 (a) (i) xylem **(1)** (ii) phloem **(1)**
(b) sieve tube elements **(1)** companion cells **(1)**
(c) Both xylem and phloem tissues form long narrow tubes. **(1)**
The cells of the xylem tissue are dead / have no organelles, and the cell walls are lined with lignin. **(1)**
The cells of the phloem are living / contain some organelles. **(1)**
There are no end plates between the cells of the xylem but there are sieve plates between the cells of the phloem. **(1)**
(d) Sugars move down the tree from the leaves to the roots during the summer. **(1)**
The ring in the bark cut through the phloem, so the sugars build up above the cut, causing the bark to bulge out. **(1)**

49. Gas exchange surfaces

1 (a) 6 cm^2 **(1)** (b) 33 cm^2 **(1)**
(c) the thinner cuboid **(1)** larger surface area **(1)**
shorter diffusion pathway / increased volume of gas can diffuse at the same time **(1)**
2 (a) large multicellular organisms **(1)**
(b) A rich blood supply can provide sufficient gases quickly enough for all the metabolic reactions to take place. **(1)**
Waste gases can be removed from the organism quickly. **(1)**
(c) Large gas spaces between spongy mesophyll cells. **(1)**
Gases enter the leaf through the stomata. **(1)**

50. The lungs

1 (a) **B (1)**
(b) Smooth muscles can contract, restricting the lumen of the bronchus **(1)** to prevent harmful gases from entering the lungs. **(1)**
(c) Blood capillaries bring carbon dioxide to the alveoli **(1)** and take oxygen away from the alveoli. **(1)**
This helps to maintain a high concentration gradient of gases at the alveoli. **(1)**
2 There will be fewer alveoli, as the elastic fibres in the alveolar wall have been broken down, causing many alveoli to merge. **(1)**
This will reduce the surface area for gas exchange. **(1)**

51. The mechanism of ventilation

1 **B (1)**
2 (a) The external intercostal muscles contract and move the ribcage upwards and outwards. **(1)**
The diaphragm contracts and becomes flatter. **(1)**
The volume of the thorax increases **(1)**
and the pressure in the thorax decreases. **(1)**
(b) The pressure in the thorax increases during expiration, above atmospheric pressure. **(1)** This forces air out of the lungs, as air moves from high pressure to low pressure. **(1)**
3 Some of the air remains in the bronchi / bronchioles / airways / dead space. **(1)** Some air remains in the alveoli as they cannot be completely flattened due to the presence of surfactant. **(1)**

52. Using a spirometer

1 **D (1)**
2 (a) 12 breaths taken during time X **(1)**
time X is 1 minute, so 12 breaths per minute **(1)**
(b) 3.75 dm^3 **(1)**
(c) The air chamber in the spirometer rises and falls as the person breathes. **(1)** As the person breathes, oxygen is removed from the spirometer and the carbon dioxide breathed out is absorbed **(1)** so the air chamber does not rise as high. **(1)**

53. Ventilation in bony fish and insects

1 **C (1)**
2 Blood flows into the secondary lamellae. **(1)**
The water flows over the secondary lamellae in the opposite direction to create a countercurrent flow. **(1)**
3 (a) A spiracle **(1)** B trachea **(1)**
C tracheole **(1)**
(b) Oxygen from the air diffuses into the insect's body cavity through the spiracles, and into long thin tubes called tracheae. **(1)**
The tracheae branch into smaller tubes called tracheoles that have an open ending inside the insect cell, filled with tracheal fluid. **(1)**
Oxygen diffuses into this fluid, and into the insect's cells. **(1)**

54. Circulatory systems

1 **A (1)**
2 (a) Oxygen diffuses into the cell across the cell surface membrane **(1)** from high oxygen concentration outside the cell to low oxygen concentration inside the cell. **(1)**
(b) Single-celled organisms have a large surface area to volume ratio. **(1)** Oxygen can diffuse to the centre of the organism quickly enough for metabolic reactions to take place. **(1)**
3 (a) Fish hearts and human hearts both have chambers / an atrium and a ventricle. **(1)**
Fish hearts have two chambers: one atrium and one ventricle. **(1)**
Human hearts have four chambers: two atria and two ventricles. **(1)**
(b) Blood moves around the body more quickly. **(1)**

55. Blood vessels

1 **D (1)**
2 (a) squamous epithelial cells **(1)**
(b) The elastic fibres allow the smooth muscles to recoil, after being stretched by the flow of blood **(1)**
so blood flow continues in the correct direction. **(1)**
3 (a) three from:
narrow lumen, to maintain high blood pressure **(1)**
folded endothelium, so lumen can get wider during blood flow **(1)**
thick walls / smooth muscle layer / layer of collagen fibres to withstand a high blood pressure **(1)**
thick layer of elastic fibres, to stretch and recoil after the flow of blood / pulse **(1)**
(b) Skeletal muscles surrounding veins compress the veins. **(1)**
Veins contain valves, so blood only flows in one direction. **(1)**

56. The formation of tissue fluid

1 (a) the pressure created by the osmotic effects of the solutes **(1)**
(b) −1.2 kPa **(1)**
(c) Fluids move from high pressure to low pressure. **(1)**
The fluid will move from inside the blood vessel to outside the blood vessel, to form tissue fluid. **(1)**

(d) The fluid, which contains high concentrations of oxygen and glucose, has moved closer to the body's cells. **(1)**
There is less distance for oxygen and glucose to diffuse into the cells and for carbon dioxide to diffuse out of the cells. **(1)**

2 (a) to identify pathogens **(1)** and engulf them by phagocytosis **(1)**
(b) Erythrocytes are too large to pass through the small gaps in the walls of the blood capillaries. **(1)**

57. The mammalian heart

1 A (right) pulmonary vein **(1)**
B right ventricle **(1)**
C (left) pulmonary artery **(1)**
D bicuspid / atrioventricular valve **(1)**

2 (a) Blood enters the left atrium from the lungs, through the pulmonary artery. **(1)**
Blood flows through the bicuspid / atrioventricular valve, into the left ventricle. **(1)**
Blood flows through the semilunar valve into the aorta and around the body. **(1)**
(b) Blood in left side has a higher concentration of oxygen **(1)** and a lower concentration of carbon dioxide. **(1)**

3 The right ventricle has less muscle than the left ventricle. The right ventricle creates enough pressure to push the blood around the lungs. The left ventricle provides a greater pressure than the right ventricle as the blood travels from the left ventricle around the body. **(3)**

58. The cardiac cycle

1 **C (1)**

2 (a) The right atrium contracts, increasing the pressure inside the atrium above the pressure inside the right ventricle. **(1)**
The atrioventricular valves are pushed open. **(1)**
Blood moves from high pressure to low pressure. **(1)**
(b) After the blood flows into the ventricles, the ventricles contract. **(1)**
This increases the pressure inside the ventricles above the pressure in the atria, causing the atrioventricular valves to be pushed closed. **(1)**

3 120 is the systolic pressure, which is the arterial pressure when the ventricles contract. **(1)**
80 is the diastolic pressure, which is the arterial pressure when the ventricles are relaxed. **(1)**

59. Control of the heart

1 **B (1)**

2 The sinoatrial node / SAN sends out a wave of excitation over both atria, causing both atria to contract. **(1)**
When the wave of excitation reaches the atrioventricular node / AVN, it passes down the Purkyne fibres in the septum, **(1)** causing the ventricles to contract. **(1)**

3 The tallest peak is the QRS complex / wave. **(1)**
It represents the contraction of the ventricles. **(1)**

60. Haemoglobin

1 About 5% is dissolved directly in the plasma. **(1)**
About 10% combines directly with haemoglobin to form carbaminohaemoglobin. **(1)**
About 85% is transported as carbonic acid. **(1)**
Carbon dioxide is combined with water by an enzyme called carbonic anhydrase, to form carbonic acid. **(1)**

2 **(a)**

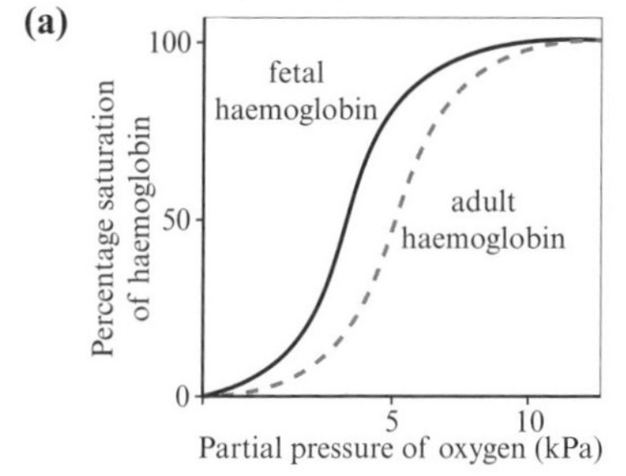

Curve should be a similar shape to the adult curve. **(1)**
Fetal curve should be to the left of the adult curve. **(1)**
(b) Tissues have low partial pressures of oxygen, and so most of the oxygen dissociates from haemoglobin. **(1)**
At high partial pressures of oxygen in the lungs, most of the oxygen associates with haemoglobin / haemoglobin is saturated with oxygen. **(1)**
The curve is not linear, because as each haem group in haemoglobin associates with oxygen, it becomes easier for the other haem groups to associate with oxygen. **(1)**

61. The plant vascular system

1 **D (1)**

2 (a) Xylem is made from cells called xylem vessel elements. **(1)**
Xylem contains lignin which gives strength to the cell walls. **(1)**
(b) The xylem is hollow so that water and dissolved minerals can pass along the narrow tube. **(1)**
The tube is narrow to allow the cohesion between the water molecules to cause capillary action. **(1)**
This allows the water to flow up the tubes. **(1)**

3 (a) sucrose **(1)**
(b) Phloem cells are living and contain some organelles, whereas xylem cells are dead. **(1)**
Phloem cells are connected by sieve plates, whereas xylem cells have no end plates between them. **(1)**
(c) Phloem companion cells use ATP to load and offload sugars from the phloem. **(1)**
Loading and offloading is carried out by active transport. **(1)**

62. Leaves, stems and roots

1 A epidermis **(1)** B xylem **(1)**
C cambium **(1)** D phloem **(1)**

2 (a) The xylem and phloem are in a vein. **(1)**
Xylem is towards the top surface of the leaf, with the phloem beneath. **(1)**
(b) osmosis **(1)**
(c) five from:
Companion cells load the sugars / sucrose onto the phloem from the leaf, which is the source during the summer. **(1)**
Hydrogen / H^+ ions are actively transported out of the companion cells, creating a concentration gradient / higher concentration of hydrogen / H^+ ions outside of the cell. **(1)**
Hydrogen / H^+ ions move back into the companion cells through a cotransporter protein by facilitated diffusion. **(1)**
As each hydrogen / H^+ ion diffuses through the co-transporter protein, it transports one sucrose molecule with it. **(1)**
Sucrose diffuses from the companion cells into the phloem through the plasmodesmata. **(1)**
The phloem transports the sugars down the plant to the roots, which in summer are the sink, where they are offloaded by the companion cells. **(1)**
Sucrose diffuses out of the phloem through the plasmodesmata and into the root cells. **(1)**
Sucrose may be actively transported out of the phloem. **(1)**

63. Transpiration

1 **B (1)**

2 (a) Water moves from the xylem to the mesophyll cells by osmosis. **(1)**
Water evaporates from the surface of the mesophyll cells and forms water vapour in the spaces between the mesophyll cells, which creates a high water vapour potential in the leaf. **(1)**
Water diffuses out of the leaf through the stomata from high water vapour potential to low water vapour potential. **(1)**
(b) Fewer stomata are open at night **(1)** as stomata close in the absence of light. **(1)**

3 A plant is cut underwater to prevent any air bubbles forming in the xylem and placed into a capillary tube. **(1)**

A bubble is introduced into the capillary tube and the movement of the bubble is measured over time. **(1)**
This is used to estimate how much water has been lost by transpiration. **(1)**

64. The transport of water

1 C **(1)**

2 Hydrogen bonds form between adjacent water molecules, forming by cohesion. **(1)**
Hydrogen bonds also form between water molecules and the xylem wall, forming by adhesion. **(1)**
Water is pushed up the xylem by the root pressure **(1)**
and pulled up the xylem by transpiration and capillary action. **(1)**

3 Many large air spaces in the leaf to keep the leaves afloat. **(1)**
Stomata are found on the upper epidermis so that they are exposed to the air **(1)** and air can diffuse in and out of the leaf. **(1)**
Aerial leaf surfaces are waterproof to prevent the blocking of stomata. **(1)**
The leaf stem has many large air spaces to allow oxygen to diffuse quickly to the roots for aerobic respiration. **(1)**

65. Translocation

1 D **(1)**

2 (a) Water moves from the xylem into the top of the phloem by osmosis, from an area of high water potential to an area of low water potential. **(1)**
This creates a high pressure at the top of the phloem, forcing the phloem sap down the phloem. **(1)**
The water potential at the top of the phloem is low because of the high sucrose concentration. **(1)**
(b) molecules made by the plant from glucose, such as amino acids or fats **(1)**

3 Two flasks of sugar solution are set up, one dilute and the other concentrated, connected by a tube. **(1)**
Each flask has a partially permeable membrane. **(1)**
Water moves across the partially permeable membrane into the high concentration sugar solution flask, because of the difference in osmotic pressure. **(1)**
The concentrated sugar solution then moves through the tube into the other flask. **(1)**

66. Exam skills

1 (a) Diaphragm contracts and becomes flatter. **(1)**
External intercostal muscles contract and move the ribcage upwards and outwards. **(1)**
The volume of the thorax increases, and the pressure inside the lungs decreases below atmospheric pressure. **(1)**
Air flows into the lungs from high pressure to low pressure. **(1)**
(b) (i) 6 breaths in 12 seconds **(1)**
$\frac{6}{12} \times 60 = 30$ breaths per minute **(1)**
(ii) vital capacity **(1)**
(iii) Make sure the spirometer is sterilised. **(1)**
Make sure the person using the spirometer is not too ill to use the machine. **(1)**

67. Types of pathogens

1 (a) A pathogen is a microorganism that causes disease. **(1)**
(b) virus **(1)**
(c) fungi **(1)**
(d) Protoctista **(1)** Fungi **(1)**

2 A **(1)**

3

Disease	Type of pathogen
influenza	virus **(1)**
ring rot	bacteria **(1)**
tomato blight	protozoan **(1)**
ringworm	fungus **(1)**

68. Transmission of pathogens

1 (a) the transmission of a pathogen from an infected organism to a healthy organism **(1)**
(b) one from: HIV / tuberculosis / cholera / anthrax **(1)**
(c) Spores from the infected plant can be carried in the air **(1)** or can be present in the soil. **(1)**

2 A **(1)**

3 (a) Fungi on the infected plant produce spores. **(1)**
An insect eating the plant would pick up the spores and carry them to an uninfected plant when it feeds. **(1)**
Insecticide kills the insects and prevents them from spreading the spores from the infected plants. **(1)**
(b) fungicide **(1)**

69. Plant defences against pathogens

1 D **(1)**

2 (a) Tyloses form in the xylem. **(1)**
They are a swelling that blocks the xylem, and so prevent pathogens from travelling any further through the plant. **(1)**
The tylose is filled with plant chemicals called terpenes. **(1)**
(b) one from: tobacco mosaic virus / fungi / bacteria / virus **(1)**

3 Both have a permanent physical barrier against infection: plants have cellulose cell walls; animals have skin. **(1)**
Both prevent entry of pathogens: plants close the stomata to prevent pathogens from entering the plant, whereas animals form blood clots around an open wound in the skin. **(1)**
Both block movement of pathogens: plants form tyloses and calloses to block movement of pathogens through the plant; animals trap pathogens in mucus and use ciliated epithelium to remove it from the body. **(1)**
Both use chemicals: plants use chemicals such as terpenes; animals use hydrochloric acid and lysozymes to kill pathogens. **(1)**

70. Animal defences against pathogens

1 (a) a defence against any pathogen **(1)**
(b) hydrochloric acid in the stomach **(1)**
lysozyme enzymes found in tears, saliva and sweat **(1)**

2 C **(1)**

3 (a) Damage to a blood vessel exposes collagen and starts the blood clotting cascade. **(1)**
Blood clotting factors and calcium ions / Ca^{2+} work together to activate soluble fibrinogen in the blood to form insoluble fibrin. **(1)**
Fibrin attaches to platelets to form platelet plugs at the site of the cut, to prevent pathogens from entering the body. **(1)**
(b) Mast cells produce the chemical histamine, which causes the blood capillaries surrounding the cut to vasodilate. Histamine also causes the blood capillaries to become more permeable. **(1)**
This means that neutrophils can more easily leave the capillary and travel to the cut site to engulf any pathogens. **(1)**

71. Phagocytes

1 (a) neutrophils **(1)** macrophages **(1)**
(b) Phagocytes engulf pathogens by the process of phagocytosis. **(1)** A vesicle / phagosome forms around the pathogen, which combines with a lysosome inside the phagocyte. **(1)**
The phagolysosome contains enzymes that digest the pathogen. **(1)**
(c) Cytokines attract phagocytes to the site of infection by a process called chemotaxis. **(1)**
The phagocytes move from a lower concentration of cytokine towards a higher concentration of cytokine. **(1)**

2 macrophage **(1)**
antigens **(1)**
on the surface to become an antigen-presenting cell. **(1)**
This activates helper T lymphocytes. **(1)**

3 B **(1)**

72. Lymphocytes

1 D **(1)**

2 (a) Clonal selection is when the specific B or T lymphocyte is activated. **(1)**
Each B and T lymphocyte can only be activated by a specific antigen. **(1)**
(b) Each T lymphocyte is activated by a specific antigen **(1)** displayed on the surface of an infected cell / macrophage / antigen-presenting cell. **(1)**
(c) killer T cell **(1)** helper T cell **(1)**
memory T cell **(1)** regulatory T cell **(1)**
(d) Helper T cells secrete cytokines / interleukins **(1)** that stimulate phagocytes to carry out phagocytosis. **(1)**

73. Immune responses

1 (a) An antibody is made of two light chains and two heavy chains, connected by disulfide bridges. **(1)**
One end of each chain has a constant region and the other end has a variable region. **(1)**
The variable region is complementary to only one antigen / the antigen binds to the variable region. **(1)**
The constant region is the same in all antibodies and can bind to the surface of the B lymphocytes. **(1)**
(b) Antibodies bind to the antigens on the surface of pathogens, **(1)**
marking the pathogen to be engulfed by phagocytes. **(1)**

2 (a) B lymphocyte memory cells **(1)**
(b) natural active immunity **(1)**
(c) The body is injected with a dead or harmless pathogen. **(1)**
The complementary lymphocytes are activated by the antigen on the pathogen's surface. **(1)**

74. Types of immunity

1 B **(1)**

2 (a) An autoimmune disease is when the immune system attacks its own body cells. **(1)**
Your body cells have antigens on the surface that identify them as belonging to your body. **(1)**
If the white blood cells stop recognising these antigens as belonging to the body, the white blood cells attack the body cells. **(1)**
(b) two from: rheumatoid arthritis **(1)**
lupus / systemic lupus erythematosus (SLE) **(1)**
type 1 diabetes **(1)**
any other autoimmune disease **(1)**

3 (a) In a person who has not been vaccinated against tetanus, there are no memory cells to produce antibodies to help fight the infection **(1)**
and there is no time to make antibodies **(1)**
so an injection of antibodies is given and these will bind to the bacteria. **(1)**
(b) artificial passive immunity **(1)**

75. The principles of vaccination

1 B **(1)**

2 (a) Herd immunity is when most of a population are vaccinated against a particular disease. **(1)**
This means that the disease is much less likely to be transmitted through the population. **(1)**
It also protects the few individuals who are not vaccinated. **(1)**
(b) A disease is more likely to move through the population. **(1)**
Individuals who are not vaccinated are more likely to catch the disease. **(1)**

3 three from:
from live pathogen that has been made harmless / attenuated **(1)**
from dead pathogen **(1)**
by removing the antigens from a pathogen **(1)**
from a similar, less harmful pathogen **(1)**
from a harmless version of a toxin / toxoid **(1)**

76. Vaccination programmes

1 C **(1)**

2 (a) An epidemic is the rapid spread of disease through a high proportion of the population. **(1)**
(b) Everyone, or almost everyone, in an area where a disease is possible is vaccinated. **(1)**
This leads to herd immunity and prevents the disease from spreading through the population. **(1)**

3 (a) three from:
old people **(1)** young children **(1)** pregnant women **(1)**
people with medical conditions / weakened immune systems **(1)**
(b) The influenza virus mutates rapidly, which alters the shape of the antigens on the surface. **(1)**
A new vaccine should be made using the new antigens. **(1)**

77. Antibiotics

1 There is a clear correlation in the graphs between increasing use and increasing resistance. The increased use of the antibiotic means that more bacteria are exposed to it. Some of these bacteria may have undergone mutation that gives them an advantage over others due to resistance to the antibiotic. The ones with the resistance mutation are selected for, replicate and pass the resistance allele on to their offspring. These antibiotic-resistant organisms become more common. The resistant allele may also be passed to other bacteria by the plasmids or conjugation, a form of sexual reproduction in bacteria. **(6)**

2 (a) the redesign or construction of new biological molecules or cells **(1)**
for example, synthetic bacteria / new enzymes / fortified food **(1)**
(b) Personalised chemotherapy drugs could be delivered to cancer cells using synthetic bacteria. **(1)**
New biological parts could be made synthetically to replace damaged human organs. **(1)**

78. Exam skills

1 (a) *Mycobacterium tuberculosis* **(1)**
(b) *Mycobacterium tuberculosis* is carried in droplets in air caused by coughing or sneezing. **(1)**
A person is infected by breathing in droplets. **(1)**
(c) antibiotics **(1)**
(d) five from:
A vaccine contains the dead or harmless pathogen / pathogen's antigen. **(1)**
Injected vaccine triggers a specific immune response by the B lymphocytes. **(1)**
Antibodies on the B lymphocyte surface that are complementary to the pathogen's antigen cause clonal selection and clonal expansion. **(1)**
Many antibodies that can bind to the antigen are made. **(1)**
The B lymphocytes produce memory cells that remember the antigen. **(1)**
Macrophages engulf pathogens that have antibodies attached to them. **(1)**
Macrophages display the pathogen's antigens on their surfaces and become antigen-presenting cells. **(1)**
Antigen-presenting cells stimulate helper T lymphocytes. **(1)**
(e) Most people in the UK are not infected with tuberculosis, so infection is unlikely. **(1)**

79. Measuring biodiversity

1 (a) Species evenness is a measure of the relative number of individuals in each species in a habitat. **(1)**
Species richness is the number of different species. **(1)**
(b) Use randomly placed quadrats. **(1)**
Count the number of individuals of each species in each quadrat. **(1)**

2 B **(1)**

3 Ensures all areas of a habitat are sampled **(1)** and can more easily identify changes in a habitat. **(1)** Rare species are less likely to be missed **(1)** which gives a better indication of the biodiversity of the habitat. **(1)** Non-random sampling is more time consuming **(1)** and so may not be possible in large habitats or difficult terrain. **(1)**

80. Sampling methods

1 Place the frame quadrat down in an area, using randomly generated coordinates. **(1)**
Record the species present in the quadrat, and the number of individuals of each species. **(1)**

2 **A (1)**

3 (a) systematic sampling **(1)**
(b) can show the change in species across an area **(1)**
can show a change in abiotic factors across an area **(1)**

81. Simpson's index of diversity

1 **B (1)**

2 (a) N (2010) = 16 **(1)**
$\left(\frac{n}{N}\right)^2$ (freshwater shrimp, 2010) = 0.016 **(1)**
$\left(\frac{n}{N}\right)^2$ (bloodworm, 2015) = 0.010 **(1)**
D (2015) = 0.734 **(1)**
(b) The Simpson's index of diversity is higher in 2015, indicating a higher level of biodiversity. **(1)**
There are more clean water species / water snails / shrimp / mayfly nymph in 2015. **(1)**
There are fewer water pollution indicator species / bloodworms / rat-tailed maggots in 2015. **(1)**
(c) The water is less polluted in 2015. **(1)**
There is more oxygen / food available / fewer predators of water snails / shrimp / mayfly nymph. **(1)**

82. Factors affecting biodiversity

1 (a) Generic diversity is the amount of genetic variation within a population. **(1)**
(b) small number of individuals / rare species **(1)**
(c) By working out the number of genes that are polymorphic.
Express this as a proportion of the total number of genes. **(1)**

2 **A (1)**

3 (a) two from: monoculture **(1)**
developing land for homes and to grow crops **(1)**
climate change **(1)**
hunting animals to extinction **(1)**
(b) proportion of polymorphic gene loci =
$\frac{\text{number of polymorphic gene loci}}{\text{total number of loci}}$ **(1)**
$\frac{300}{2800}$ **(1)** = 0.107 **(1)**
(c) The population has low genetic diversity. **(1)**

83. Maintaining biodiversity

1 (a) conserving a plant or animal species in its natural environment **(1)**
(b) two from:
Biodiversity in that ecosystem is protected. **(1)**
It permanently protects significant elements of natural and cultural heritage. **(1)**
It allows management of these areas to ensure that ecological integrity is maintained, improved or restored. **(1)**

2 **D (1)**

3 (a) The Convention on International Trade in Endangered Species (CITES) prevents the trade of endangered wild plants, animals and animal parts.
The Rio Convention on Biological Diversity promotes worldwide cooperation towards sustainable development and increasing biodiversity. **(1)**
The Countryside Stewardship Scheme aims to increase biodiversity across the UK. **(1)**
(b) If endangered animals are left in their natural habitat there will be a greater number of breeding pairs. **(1)**
This means that there are more possible alleles available, and therefore more genetic variation. **(1)**

84. Classification

1 **B (1)**

2

Domain	Eukarya
Kingdom	Animalia
Phylum (1)	Chordata
Class	Mammalia
Order	Carnivora
Family (1)	Canidae
Genus	**Canis (1)**
Species	**lupus (1)**

3 Compare DNA of different species. **(1)**
The more similar the DNA, the more closely related the species. **(1)**
Compare the amino acid sequence of a common protein / cytochrome c. **(1)**
The more similar the amino acid sequence, the more closely related the species. **(1)**

85. The five kingdoms

1 **C (1)**

2 (a) An organism that makes its own organic compounds from inorganic substrates, using light or chemical energy. **(1)**
(b) all plants **(1)**
some protoctists / prokaryotes **(1)**
(c) observation of features, such as anatomy
observation of behaviour **(1)**
examination of DNA sequence **(1)**
examination of sequence of biological molecule / RNA / amino acids in proteins **(1)**

3 (a) Archaea, Bacteria (or Eubacteria), Eukarya **(1)**
(b) All organisms in the Archaea and Eubacteria domains are prokaryotes. **(1)**
Archaea and Eubacteria have different ribosomal RNA. **(1)**

86. Types of variation

1 (a) Characteristics have a range and can be placed on a continuum. **(1)**
(b) height / weight **(1)**
(c) two or more different genes involved / polygenic **(1)**
environmental influences involved **(1)**

2 **B (1)**

3 The critical value at $p = 0.05$ at 1 degree of freedom is 12.7. **(1)**
The t-test value is less than 12.7. **(1)**
Therefore, the reaction times are not significantly quicker than without caffeine. The differences could be due to chance. **(1)**

87. Evolution by natural selection

1 Dolphins and sharks are examples of convergent species. **(1)**
Both species experience the same selective pressures and have the same adaptations. **(1)**
Mutations leading to similar bodies were selected for in both creatures. **(1)**

2 (a) If these two species interbreed **(1)**, they do not produce fertile offspring. **(1)**
(b) *Cephalorhynchus* **(1)**
(c) Selection pressure from the ocean habitat acted on the dolphins. Those with a chance mutation had an advantageous allele for smooth bodies. Those individuals with this advantageous allele would survive and breed. The idea of an advantageous allele is passed on to future generations. This means the frequency of advantageous alleles in the population increased. **(6)**

88. Evidence for evolution

1 (a) Use the fossil record to compare the skeleton of the extinct species with the wild horse skeleton. **(1)**
Use preserved species to compare anatomy of both species. **(1)**
Study proteins and compare the amino acid sequences between species. **(1)**
(b) Soft tissue / hair / fur will decay, unless preserved in ice / peat. **(1)**
DNA will not always be able to be extracted intact. **(1)**
Biological molecules / RNA will not always be able to be extracted. **(1)**
It is not possible unable to observe behaviour. **(1)**

2 **C (1)**

89. Exam skills

1 (a) the number of different species that are found in a habitat **(1)**
(b) Use a frame / point quadrat. **(1)**
Place quadrats randomly in each area being studied. **(1)**
(c) The area has high biodiversity. **(1)**
There are a lot of different species and there are a lot of individuals of each species. **(1)**
(d) The area became less polluted. **(1)**
The habitat has been conserved / protection of the habitat. **(1)**
(e) Building houses / farming / transport increases pollution. **(1)**
An increase in global warming could lead to a reduction of habitats. **(1)**

90. The need for communication

1 nervous **(1)** hormonal **(1)**

2 (a) Stimulus is a change in the environment that brings about a response. **(1)**
(b) Response is a change in the behaviour or physiology of an organism resulting from a stimulus. **(1)**

3 **A (1)**

4 exchange of substances **(1)**
medium for solutes and gases, e.g. CO_2 and oxygen **(1)**
removal of metabolic waste **(1)**

5 enables cells to communicate with each other across the body **(1)**
enables rapid communication **(1)**
enables short-term or long-term responses **(1)**

91. Principles of homeostasis

1

	Nervous	Hormonal	
body temperature	✓		(1)
blood glucose concentration		✓	(1)
water potential of the blood		✓	(1)
carbon dioxide concentration of the blood	✓		(1)

2 Negative feedback is the mechanism that reverses a change, **(1)** bringing the system back to the optimum steady state. **(1)**

3 increase in temperature leads to an increase in enzyme reaction rate **(1)**
enzymes have optimum temperatures **(1)**
increase above optimum leads to enzyme denaturing / change of shape of active site / breaking of bonds (ionic, covalent, hydrogen) / change in 3D shape **(1)**
rate drops rapidly as enzyme is denatured **(1)**

92. Temperature control in endotherms

1 Endotherms use internal sources of heat. **(1)**
They upregulate their metabolism to stay warm if the external temperature falls. **(1)**

2 **D (1)**

3 Peripheral temperature receptors
send impulses to the thermoregulatory centre in the hypothalamus (if body temperature is too high or too low).
Nervous and hormonal systems carry signals to skin / liver / muscles.
vasodilation or vasoconstriction of arterioles near to the surface of the skin. Liver respiration levels increase or decrease / skeletal muscles reduce contractions or shiver.
Temperature returns to optimum level. **(6)**

4 Arterioles have smooth muscle tissue in their walls which can contract / relax **(1)** but capillaries do not. **(1)**

93. Temperature control in ectotherms

1 **A (1)**

2 orientate body towards the Sun **(1)**
expose maximum surface area for more heat absorption **(1)**

3 two from: require less food **(1)**
less time spent hunting **(1)**
more of the energy gained from food can be used for growth **(1)**

4 Cellular activity such as respiration releases energy in the form of heat. **(1)**
Enzymes function at an optimum temperature, **(1)**
and may become denatured by an increase in temperature and so become less effective. **(1)**

5 They do not need energy from food to keep warm. **(1)**

94. Excretion

1 Removal **(1)** of metabolic waste (from the body). **(1)**

2 (a) urea **(1)** carbon dioxide **(1)**
(b) Excessive consumption of protein leads to a high concentration of amino acids. **(1)**
Deamination produces ammonia, which combines with carbon dioxide to produce urea (or suitable chemical equation / $CO_2 + 2NH_3 \rightleftharpoons (NH_2)_2CO + H_2O$). **(1)**

3 Respiration produces carbon dioxide.
Carbon dioxide and water react together to produce carbonic acid / H_2CO_3. **(1)**
Carbonic acid dissociates to release hydrogen ions and hydrogen carbonate ions / $H_2CO_3 \rightarrow H^+ + HCO_3^-$ **(1)**
This reaction takes place in red blood cells and is catalysed by the enzyme carbonic anhydrase. **(1)**

4 **B (1)**

95. The liver – structure and function

1 hepatic artery **(1)** hepatic portal vein **(1)**

2 (a) The hepatic portal vein contains blood entering the liver having come from the digestive system. **(1)**
This blood is richer in nutrients, e.g. glucose (sugars) / amino acids, **(1)**
than blood in the hepatic vein, which is blood leaving the liver. **(1)**
(b) small intestine / duodenum **(1)**
(c) breakdown and recycling of old red blood cells **(1)**

3 **A (1)**

4 mitochondria **(1)**
because they produce a lot of ATP **(1)**

96. The kidney – structure and function

1 suitable labelled diagram **(3)**

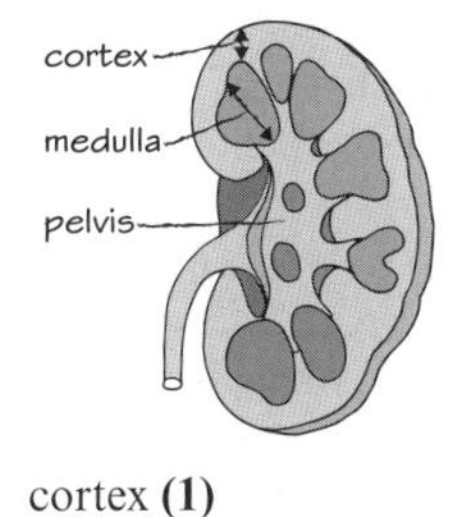

2 cortex **(1)**

3 glomerulus **(1)** Bowman's capsule **(1)**

4 **B (1)**

5 blood in renal artery: higher in oxygen concentration **(1)**
higher in urea **(1)** higher in sodium ions / sodium salts **(1)**

Could also be expressed as
blood in renal vein: lower in oxygen **(1)** lower in urea **(1)** lower in sodium ions / sodium salts **(1)**

97. Osmoregulation

1 Lumen of afferent arteriole is wider **(1)** than the lumen of the efferent arteriole. **(1)**

2

Substance	Blood	Glomerular filtrate	Urine	
red blood cells	✓	✗	✗	
amino acids	✓	✓	✗	**(1)**
glucose	✓	✓	✗	**(1)**
urea	✓	✓	✓	**(1)**

3 There is selective reabsorption in the PCT of amino acids and glucose **(1)**
by active transport and cotransport. **(1)**
Water and sodium ions are reabsorbed in the PCT and loop of Henle **(1)**
by the hairpin countercurrent multiplier effect. **(1)**

4 active transport **(1)** of sodium and chloride ions **(1)**

98. Kidney failure and urine testing

1 using a partially permeable membrane **(1)**
to filter urea (and other waste products) out of the blood **(1)**
while balancing water content and maintaining glucose and amino acid concentrations **(1)**

2

	Go up	Go down	Stay the same	
urea		✓		**(1)**
glucose			✓	**(1)**
amino acids			✓	**(1)**
salts		✓		**(1)**

3
1. hCG (human Chorionic Gonadotrophin) (1)
2. Urine (1)
3. Mobile antibodies, tagged with a blue bead (1)
4. The hCG antibody complex (1)
5. Two blue lines (1)
6. Blue beads are attached to the hcG/mobile antibody complex and they bind to the immobilised antibodies and move no further.

99. Exam skills

1 Heat energy from blood in the capillaries is absorbed by sweat and used to break hydrogen bonds in water, creating water vapour. **(1)**
The energy needed to turn water into water vapour is called the latent heat of vaporisation. **(1)** This takes heat from the body. **(1)**

2 D, the hypothalamus **(2)**

3 (a) 0.96% **(1)**
(b) Drugs like amphetamines can cause long-term adverse effects to the user. **(1)** In addition, some would say drug use would confer unfair advantage on the users. **(1)**
On the other hand, many would maintain that individuals have the right to make their own choices and a responsibility to live with the consequences, so a ban is unnecessary. **(1)**

100. Sensory receptors

1 pressure causes the corpuscle to change shape (1)
When the membrane stretch the sodium ion channels widen **(1)**
Sodium ions diffuse into the neurone **(1)**
The membrane becomes depolarized **(1)**
A generator potential is established **(1)**
An action potential is initiated and moves along the sensory neurone **(1)**

2 two from: carotid artery **(1)** brain / medulla oblongata **(1)** aorta **(1)** taste buds on the tongue **(1)** olfactory receptors in the nose **(1)**

3 **D (1)**

4 pressure **(1)** movement **(1)**

5 electrical **(1)**

101. Types of neurone

1 (a) X = axon **(1)**
Y = Schwann cell cytoplasm **(1)**
Z = nucleus **(1)**
(b) node of Ranvier **(1)**
(c) speeds it up **(1)**

2

	Sensory	Motor	Relay	
has a cell body in the CNS but an axon in the peripheral nervous system (PNS)		✓		**(1)**
whole cell found in the CNS			✓	**(1)**
carries an action potential from the CNS to an effector organ		✓		**(1)**
has a large cell body at one end of the cell		✓	✓	
has a cell body to one side of the cell which is **not** in the CNS	✓			**(1)**
has the longest axon	✓			**(1)**
is most likely to form the largest number of synapses with other neurones			✓	**(1)**

3 Action potentials / nerve impulses would travel too slowly. **(1)**
Response times between the CNS and effector organs would be too slow. **(1)**

102. Action potentials and impulse transmission

1 Correct order for **3** marks; **2** marks for 5 out of 7 correct
C B D F A E G

2 A higher frequency of action potentials **(1)**
means a more intense stimulus. **(1)**

3 At rest sodium/potassium ion pumps use ATP **(1)**
to pump 3 sodium ions out of the cell for every 2 potassium ions pumped in. **(1)**
The plasma membrane is more permeable to potassium ions / has more potassium channels than sodium ions and potassium ions diffuse out again **(1)**
but the inside of the cell remains negatively charged with respect to the outside of the cell. **(1)**

4 diagram should show a greater number of positive charges on the inside of the axon relative to the outside of the axon **(2)**

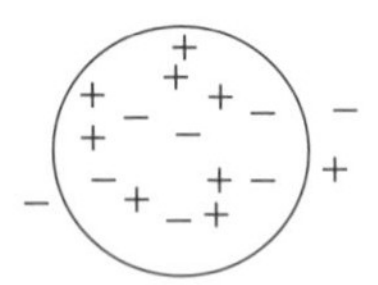

103. Structure and roles of synapses

1 Neurotransmitters are chemical substances that act as signalling molecules, released from the pre-synaptic bulge **(1)**
to diffuse across a synapse **(1)**
to bind with receptors on post-synaptic membranes. **(1)**

2 D, B, A, C (**2** marks for all four correct, **1** mark for first and last point correct)

3 Acetylcholinesterase breaks down acetylcholine so that no further action potentials are triggered on the post-synaptic membrane. **(1)**
If acetylcholinesterase was not present, this would lead to acetylcholine not being broken down. **(1)**
Continued stimulus of the muscle may lead to permanent contraction and possible paralysis. **(1)**

4 temporal summation, spatial summation **(1)**
Temporal summation occurs with multiple action potentials carried along one neurone sequentially. **(1)**
Spatial summation occurs with multiple action potentials along multiple neurones, converging at a synapse. **(1)**

104. Endocrine communications

1 A (1)
2 An endocrine gland secretes hormones into the blood. **(1)** An exocrine gland secretes other substances into ducts. **(1)** An example of an endocrine gland is the pancreas, secreting insulin. **(1)** An example of an exocrine gland is the salivary glands, secreting saliva. **(1)**
3 (a) collecting duct of the nephron in the kidney **(1)**
(b) in the blood **(1)**
4 **B (1)**
5 nervous is quicker and hormonal is slower **(1)**
nervous is electrical and hormonal is chemical **(1)**
nervous is short acting and hormone is longer acting **(1)**
nervous affects a small number of target cells and hormonal acts on a wider range of target cells **(1)**

105. Endocrine tissues

1 Adrenaline binds to a complementary receptor, causing a conformation change in the receptor that leads to an intracellular signal, e.g. second messenger / cAMP activation. **(1)**
2 When adrenaline binds the receptor protein changes conformation and the intracellular part activates the enzyme adenyl cyclase. **(1)**
Adenyl cyclase converts ATP **(1)**
to cAMP. **(1)**
3 It would let more light reach the retina / photoreceptor cells **(1)** so more detail about the surroundings is available. **(1)**
4 (a) the medulla **(1)**
(b) two from: glucocorticoids, mineralocorticoids and androgens **(2)**
5 (a) The endocrine tissue is in islets containing α cells and β cells. **(1)**
(b) one of:
enzymes **(1)** lipase **(1)**
any pancreatic enzymes **(1)** amylase **(1)**
protease (trypsinogen) **(1)**

106. Regulation of blood glucose

1 (a) islets of Langerhans in the pancreas **(1)**
(b) Insulin binds to receptors on cells (liver and muscle) **(1)**
more glucose is taken up by these cells **(1)**
and stored as glycogen. **(1)**
2 **A (1)**
3 Glucose is metabolised to produce ATP, which causes potassium ion channels to close. **(1)**
Potassium ions accumulate and this alters the potential difference across the cell membrane, **(1)**
which causes voltage-gated calcium ion channels to open and calcium ions diffuse into the cell down a concentration gradient. **(1)**
The calcium ions cause vesicles containing insulin to move towards the membrane of the cell. **(1)**
The insulin vesicles fuse with the cell membrane, releasing insulin by exocytosis. **(1)**
4 hypoglycaemia **(1)**

107. Diabetes mellitus

1 They cannot regulate their blood glucose levels. **(1)**
2 lifestyle questions: ask about diet, how much sugar / carbohydrate does he eat daily? **(1)**
family questions: is there a family history of diabetes? **(1)**
test: urine for glucose **(1)**
3 The immune system (T-cells) might incorrectly identify β-cells as foreign cells, initiating an immune response. **(1)**
The immune response could result in the destruction of the β-cells, therefore meaning an inability to make insulin. **(1)**
4 **A (1)**
5 The following are risk factors, and these are increasing in the general population: increasing incidences of obesity due to high carbohydrate / fat diets **(1)** increasingly sedentary lifestyles / lack of physical exercise. **(1)**

108. Plant responses to the environment

1 two from: phototropism **(1)** geotropism **(1)** thigmotropism **(1)** chemotropism **(1)**
2 roots produce abscisic acid (ABA)**(1)**, ABA travels to the leaves of the plant **(1)**, ABA binds to receptors on guard cells **(1)**, This brings about changes in the ionic concentration in guard cells **(1)**, which affects their water potential/turgidity **(1)**, the guard cells close the stomata. **(1)**
This causes the stomata to close, reducing water loss. **(1)**
This can lead to a lack of gas exchange and stop photosynthesis occurring. **(1)**
3 gibberellin **(1)**
4 **B (1)**
5 Digestion / breakdown of starch / endosperm **(1)**
by enzymes / amylase **(1)**
produces sugars / glucose / maltose that is used for respiration. **(1)**
6 nastic response **(1)**

109. Controlling plant growth

1 branching / sideways / lateral growth **(1)**
2 (a) auxin **(1)**
(b) Trimming hedge tops will remove the apical buds, causing abscisic acid levels to drop **(1)**
hedge plants branch out sideways, appearing bushier. **(1)**
3 **A (1)**
4 **C (1)**
5 genotype should show homozygous recessive form (le le) **(1)**

110. Plant responses

1 phototropism **(1)**
2 provides anchorage supporting the plant **(1)**
increases the likelihood of the plant's roots reaching water **(1)**
3 The seedlings would begin to grow towards the light; Auxin is produced in the tip of the growing shoot (coleoptile); The light from the left causes auxin to move laterally to the right hand side of the shoot; This stimulates cell elongation on the shaded side; Cells on the shaded side elongate more rapidly than those on the lit side **(6)**
4 **A (1)**

111. Commercial use of plant hormones

1 Broad-leaved weeds have a large surface area **(1)**
so more auxin is absorbed by broad-leaved weeds. **(1)**
Auxin causes rapid / increase in growth of leaves and the roots cannot support growth with the supply of nutrients. **(1)**
2 (a) ethene **(1)**
(b) one from: sweeten **(1)** colour change (green to yellow) **(1)** soften **(1)**
(c) Picking cherries while unripe reduces labour costs. **(1)**
Cherries are easier to transport while unripe. **(1)**
Then use hormone to ripen the fruit after picking and transport. **(1)**
3 Spray the crop with a chemical which inhibits plant growth. **(1)**
Use an inhibitor of gibberellin synthesis. **(1)**

112. Mammalian nervous system

1 brain **(1)** spinal cord **(1)**
2 **C (1)**
3 Somatic: motor neurones carry impulses from the CNS to skeletal muscles, **(1)**
which are under voluntary control. **(1)**
Autonomic: motor neurones carry impulses from the CNS to cardiac muscle, to smooth muscle in the gut wall and to glands. **(1)**
All these are involuntary. **(1)**

4

	Sympathetic	Parasympathetic	
decreased heart rate		✓	(1)
dilation of pupils	✓		(1)
increased blood flow to the brain	✓		(1)
increased blood flow to the digestive system		✓	(1)
increased sexual arousal		✓	(1)
decreased ventilation		✓	(1)

5 sympathetic **(1)**

113. The brain

1 heart rate **(1)** circulation and blood pressure **(1)**
ventilation rate **(1)**
2 cerebrum **(1)**
3 **B (1)**
4 The cerebellum is involved with fine control of muscular movements, such as manipulating tools or maintaining body balance. **(1)**
Such activities often require learning, but once learned involve unconscious control. **(1)**
5 Sensory input from temperature receptors **(1)** and osmoreceptors **(1)**
is received by the hypothalamus and this leads to the initiation of nervous responses that regulate body temperature **(1)**
and blood water potential. **(1)**

114. Reflex actions

1 (a) involuntary / does not involve processing by the brain **(1)**
nearly instantaneous / very quick movement in response to a stimulus. **(1)**
(b) Sensory (1), Relay (1), Motor (1).
2 **A (1)**
3 Following stimulation when the muscles at the front of the thigh are stretched, a nerve impulse passes along a sensory neurone in the upper leg towards the spinal cord. **(1)**
The impulse passes through an interneurone in the spinal cord **(1)**
before passing along a motor neurone which terminates at a neuromuscular junction in the thigh muscle, causing contraction of the muscle. **(1)**
4 suckling **(1)** grasping **(1)** rooting **(1)**

115. Coordination of responses

1 Adrenaline increases the level of glucose in the blood **(1)**
increases the ventilation rate **(1)**
increases the heart rate **(1)**
causes arterioles to the muscle to dilate. **(1)**
2 Adrenaline binds to a receptor on the surface of a cell causing a response inside the cell. **(1)**
cAMP acts on enzymes inside the cell bringing about a response. **(1)**
3 **B (1)**
4 Adrenaline travels through the blood. **(1)**
Blood reaches all cells in the body, where the adrenaline binds with receptors on the surface of different tissue types. **(1)**

116. Controlling heart rate

1 sinoatrial node (SAN) **(1)**
2 (a) chemoreceptor **(1)** stretch receptor **(1)**
(b) chemoreceptors in carotid artery/aorta/brain **(1)**
stretch receptors in muscles/carotid sinus **(1)**
3 **C (1)**
4 Stretch receptors in the muscle and chemoreceptors in the aorta, brain or carotid arteries **(1)**
send signals via sensory neurones to the medulla oblongata. **(1)**
An increase in impulses sent down the accelerator nerve increases the heart rate. **(1)**
An increase in impulses sent down the vagus nerves decreases the heart rate. **(1)**

117. Muscle structure and function

1 thin filaments = actin **(1)** thick filaments = myosin **(1)**
2 (a) will increase **(1)** (c) will increase **(1)**
(b) will increase **(1)**
3 an arrow pointing to any point where the thick and thin filaments overlap **(1)**
4 (a) When a muscle generates its own excitation for contraction / initiates its own beat at regular intervals. **(1)**
(b) heart muscle (atrial, ventricular) **(1)**
5 within arteries / arterioles / intestines / uterus **(1)**

118. Muscular contraction

1 At the synaptic knob: vesicles of neurotransmitter fuse with the pre-synaptic membrane **(1)**
and neurotransmitter is released into the synaptic cleft. **(1)**
On the muscle fibre membrane: neurotransmitter binds to receptors **(1)**
causing depolarisation (of the membrane) **(1)**
and the depolarisation wave travels down the T-tubules. **(1)**
Within the muscle fibre: T-tubule depolarisation leads to the release of Ca^{2+} from the sarcoplasmic reticulum. **(1)**
Ca^{2+} binds to troponin **(1)**
and a power stroke/contraction occurs. **(1)**
2 level of contraction increases / becomes more powerful **(1)**
such that tetanus can occur. **(1)**
3 bind to receptors in place of the neurotransmitter **(1)**
prevent release of the neurotransmitter from the pre-synaptic membrane **(1)** prevent the breakdown of the neurotransmitter once it is bound to the receptor **(1)**

119. Exam skills

1 (a) Electrical activity travels through the atria and reaches the AVN. **(1)**
The atria contract, whereby their volume decreases **(1)**
and pressure rises, forcing blood into the ventricles. **(1)**
(b) 60 ÷ 0.75 **(1)** 80 **(1)**
2 **B (1)**
3 **C (1)**

120. Photosynthesis and respiration

1 $6H_2O + 6CO_2$ **(1)**
$\rightarrow C_6H_{12}O_6 + 6O_2$ **(1)**
2 chloroplast **(1)**
3 Autotrophs photosynthesise and respire **(1)**
while heterotrophs respire only. **(1)**
4 both contain DNA **(1)**
similar size **(1)**
5 Photosynthesis produces sugars. **(1)**
Early eukaryotes could use sugars to respire and so grow more quickly / move around. **(1)**
6 Glucose moves out of chloroplasts via transport proteins, using facilitated diffusion. **(1)**
Oxygen diffuses through the chloroplast membranes. **(1)**

121. Photosystems, pigments and thin layer chromatography

1 (a) W ribosome **(1)** Y granum / thylakoid **(1)**
X stroma **(1)** Z intergranal lamella **(1)**
(b) granum / thylakoid membrane (Y) **(1)**
(c) membrane sacs with large surface area for pigments / antennae complexes to capture light **(1)**
(d) X contains the enzymes needed for the light-independent reaction (LIR) stage of photosynthesis. **(1)**
2 A photosystem consists of a funnel-shaped **(1)**
light-harvesting cluster of photosynthetic pigments **(1)**
held in place in the thylakoid membrane by proteins. **(1)**

3

	Chlorophyll a	Chlorophyll b	Carotenoids	
found in two forms: P700 and P680	✓			(1)
reflects orange and yellow light			✓	(1)
contains a porphyrin ring	✓	✓		(1)
appears blue-green		✓		(1)
accessory pigments		✓	✓	(1)

122. Light-dependent stage

1 (a) photolysis **(1)**
(b) some used for aerobic respiration **(1)**
rest diffuses out of the plant via stomata **(1)**
(c) in photosystem II, at the thylakoid membrane **(1)**
2 **C (1)**
3 Cyclic photophosphorylation involves only photosystem I while non-cyclic photophosphorylation involves photosystem I and photosystem II. **(1)**
There is no photosynthesis in cyclic photophosphorylation. **(1)**
Only ATP is produced by cyclic photophosphorylation **(1)** but reduced NADP and O_2 and ATP are produced by non-cyclic photophosphorylation. **(1)**
Electrons return to photosystem I in cyclic photophosphorylation. **(1)**

123. Light-independent stage

1 Calvin cycle **(1)**
2 (a) X is CO_2 **(1)**
Y is glycerate-3-phosphate (GP) **(1)**
(b) triose phosphate **(1)**
(c) RuBisCO / ribulose bisphosphate carboxylase-oxygenase **(1)**
3 **C (1)**
4 Hydrogen ions are pumped **(1)**
from the stroma into the thylakoid space. **(1)**
5 reduced NADP **(1)**
ATP **(1)**

124. Factors affecting photosynthesis

1 The limiting factor is the factor whose magnitude limits the rate of a natural process. **(1)**
2 X: as the light intensity increases so does the rate of photosynthesis. **(1)**
Light intensity is the limiting factor. **(1)**
Y: despite further increases in light intensity there is no additional increase in the rate of photosynthesis. **(1)**
Some other factor, such as carbon dioxide concentration or temperature, has become the limiting factor. **(1)**
3 CO_2: carbon-based fuels could be burnt to increase CO_2 concentration. **(1)**
Light: glass in the walls and roof of greenhouses allows light to reach plants. **(1)**
Light banks / artificial light can be used at night. **(1)**
Temperature: can be maintained through the use of heaters, maintaining the optimum temperature. **(1)**
water/mineral levels can be regulated **(1)**
A greenhouses restrict access of pests. **(1)**

125. The need for cellular respiration

1 The process whereby energy stored in complex organic molecules, e.g. proteins and carbohydrates, is used to make ATP. **(1)**
It occurs in living cells. **(1)**
2 Metabolic reactions that build large molecules are anabolic. **(1)**
Those that break large molecules into smaller ones are catabolic. **(1)**
3 three from: exocytosis **(1)** endocytosis **(1)** protein synthesis **(1)**
DNA replication **(1)** movement **(1)**
activation of chemicals (for example, phosphorylation of glucose in glycolysis or substrate phosphorylation in the Krebs cycle) **(1)**
4 **B (1)**
5 three phosphate groups **(1)**
one ribose sugar **(1)**
adenine / nitrogenous base **(1)**

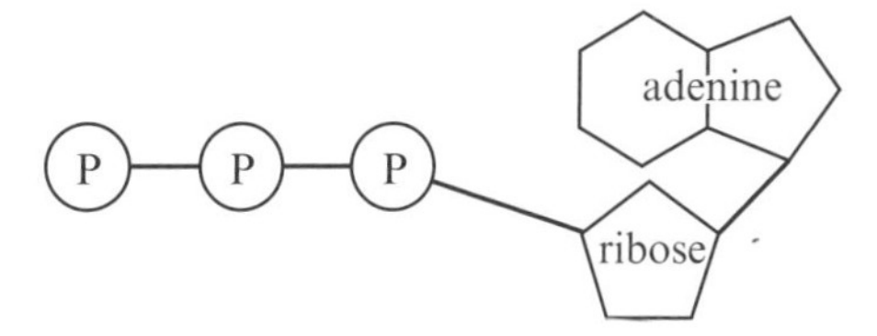

126. Glycolysis

1 occurs in cytoplasm / cytosol **(1)**
2 (a) stage X **(1)**
(b) stage Y **(1)**
(c) 4 **(1)**
3 reduced NAD **(1)**
4 **D (1)**
5 Oxidation reactions involve the loss of electrons while reduction reactions involve the addition of electrons. **(1)**
In glycolysis, the coenzyme NAD becomes reduced as the substrate (glycolytic intermediate) becomes oxidised, i.e. the reactions are coupled. **(1)**

127. Structure of the mitochondrion

1 X DNA (naked) **(1)**
Y matrix **(1)**
Z cristae / inner mitochondria membrane **(1)**
2 via (protein) carriers **(1)**
utilising active transport **(1)**
3 **A (1)**
4 impermeable to most small ions and to proteins **(1)**
folded into cristae to give large surface area **(1)**
embedded with electron carriers **(1)**
embedded with ATP synthase enzymes **(1)**

128. Link reaction and the Krebs cycle

1 matrix of the mitochondria **(1)**
2 **D (1)**
3 (a) X is oxaloacetate **(1** Y is citrate **(1)**
(b) 6 **(1)**
(c) between Y and α-ketoglutarate **(1)**
between α-ketoglutarate and X **(1)**
4

	No. made in link reaction	No. made in Krebs cycle	
reduced NAD	2	6	(1)
reduced FAD	0	2	(1)
ATP	0	2	(1)
CO_2	2	4	(1)

129. Oxidative phosphorylation

1 inner mitochondrial membrane / cristae **(1)**
2 (a) glycolysis **(1)** Krebs cycle **(1)**
(b) Glycolysis produces 4 ATP, but uses 2, so net production is 2. **(1)**
Krebs cycle produces 2 ATP. **(1)**
3 **A (1)**
4 The enzyme combines $4H^+$ **(1)**
and 4 electrons **(1)**
with O_2 to make water ($2H_2O$). **(1)**

130. Anaerobic respiration in eukaryotes

1 **A (1)**

2 (a) pyruvate **(1)** (b) lactate dehydrogenase **(1)**

3 (a) normally carried away from the muscle by the blood to the liver **(1)**
where it is converted back into pyruvate **(1)**
(b) It would cause muscle fatigue, reducing the pH / increasing the level of acidity and reducing enzyme activity in the muscles. **(1)**

4 (a) ethanal **(1)** (c) CO_2 **(1)**
(b) ethanol **(1)**

5

	Yeast	Mammals	Both	
Ethanal is the hydrogen acceptor.	✓			(1)
Carbon dioxide is produced.	✓			(1)
Reduced NAD is reoxidised.			✓	(1)
Lactate is made.		✓		(1)
Enzymes are required.			✓	(1)

131. Energy values of different respiratory substrates

1 an organic substance that can be used for respiration **(1)**

2 (a)

Substrate	Mean energy value (given in kJ g^{-1})	
carbohydrate	15.8	(1)
lipid	39.4	(1)
protein	17.0	(1)

(b) The more protons / hydrogen atoms available **(1)**
the more chemiosmosis / ATP production. **(1)**

3 **B (1)**

4 The volumes of oxygen used and carbon dioxide produced were equal. **(1)**

132. Factors affecting respiration

1 Diffusion increases with a temperature increase as collisions have more kinetic energy and are more successful. **(1)**
Enzyme-controlled reactions initially speed up with temperature increase, but high temperatures will cause enzymes to denature. **(1)**
Membranes will start to break down at higher temperatures. **(1)**
The rate of respiration will rapidly decrease when temperature increases above the optimum. **(1)**

2 **A (1)**

3 (a) oxygen uptake **(1)**
(b) stop watch **(1)**
(c) $cm^3\ min^{-1}$ **(1)**

4 The only volume change within the respirometer is due to the volume of oxygen absorbed by the organism. **(1)**

133. Exam skills

1 A is oxygen. **(1)**
B is triose phosphate/glycerate 3-phosphate. **(1)**

2 (a) X is ATPase. **(1)**
(b) The gradient is from the intermembrane space into the matrix **(1)**
so needs to be a high concentration in the space. **(1)**
H^+ ions from reduced NAD are pumped into the intermembrane space **(1)**
requiring energy derived from movement of electrons along ETC on the cristae. **(1)**

3 Runner's oxygen requirement in dm^3 per minute is $60 \times 0.3\,dm^3$, which is 18 $dm^3\ min^{-1}$. **(1)**
The runner's maximum oxygen intake is 3.5 $dm^3\ min^{-1}$ and so the rate at which they will incur their oxygen debt is $18 - 3.5\,dm^3 = 14.5\,dm^3\ min^{-1}$ **(1)**
They will incur their full oxygen debt in 1 minute. **(1)**
The runner will have run $60 \times 5 = 300$ m in that time. **(1)**

134. Gene mutation

1 a change in the DNA sequence of a gene **(1)**

2 insertion mutation **(1)**

3 (a) The reading frame of the gene sequence has been shifted by one nucleotide, causing a frameshift mutation. **(1)**
This alters the triplet codes for the amino acids. **(1)**
In this example, the third triplet has changed from GGG to TGG, which codes for a different amino acid. **(1)**
The fourth triplet has changed from TAG to GTA. **(1)**
The amino acid sequence from the point of the frameshift will be different. **(1)**
(b) As the DNA codes for an enzyme, then the active site could be a different shape and would no longer be complementary to the substrate. **(1)**
The protein may be shorter and no longer have an active site or binding site. **(1)**
The protein might be less effective or no longer functional. **(1)**

135. Gene control

1 A structural gene codes for a protein **(1)**
whereas a regulatory gene controls whether or not a structural gene is expressed. **(1)**

2 (a) In the absence of lactose, the regulatory gene produces a repressor protein that binds to the operator. **(1)**
This prevents RNA polymerase from moving from the promoter and over the operator in order to transcribe genes Z and Y. **(1)** Z and Y are not expressed. **(1)**
(b) Lactose binds to the repressor protein and changes its shape. **(1)**
The repressor protein can no longer bind to the operator sequence so RNA polymerase is able to transcribe genes Z and Y. **(1)**
(c) The organism does not always find itself in the presence of lactose. **(1)**
It takes energy to produce lactase, the enzyme needed to utilise lactose as an energy source. **(1)**
The organism saves energy by only producing lactase in the presence of lactose. **(1)**

136. Homeobox genes

1 (a) a homeotic or regulatory gene that contains a homeobox sequence **(1)**
and controls the development of the body plan **(1)**
(b) Homeobox genes are arranged in clusters. **(1)**
Groups of homeobox genes occupy the same position in different clusters. **(1)**
(c) Homeobox genes code for transcription factors. **(1)**
Transcription factors bind to the DNA with a DNA-binding domain / homeodomain. **(1)**
Transcription factors control the expression of other genes **(1)** by switching genes on and off. **(1)**

2 (a) The homeobox genes in all species have a very similar gene sequence and are located in the same positions. **(1)**
(b) Homeobox genes are very important; a mutation in them would be very likely to alter the body plan / affect many other genes. **(1)** A mutation in a homeobox gene is likely to be selected against / lethal. **(1)**

137. Mitosis and apoptosis

1 (a) growth of the organism **(1)**
repair of damaged tissue **(1)**
(b) prophase
metaphase **(1)**
anaphase **(1)**
telophase **(1)**

2 (a) Enzymes break down the cytoskeleton. **(1)**
The nucleus breaks down. **(1)**

Bulges / blebs appear in the plasma membrane. **(1)**
Cell breaks down into vesicles / apoptotic bodies. **(1)**
(b) engulfed by phagocytes / phagocytosis **(1)**

3 Chromosomes lined up in the centre of the cell. **(1)** Each chromosome attached to a spindle fibre at the centromere. **(1)**

138. Variation

1 (a) variation where the trait varies by very small amounts between one group and the next **(1)**
(b) (i) genes **(1)** (ii) environment **(1)**
(c) discontinuous **(1)**

2 (a) Sexual reproduction leads to random fertilisation. **(1)**
Independent assortment of chromosomes takes place. **(1)**
There is crossing over between homologous chromosomes. **(1)**
There can be mutations in the DNA. **(1)**
Natural selection leads to changes in allele frequency. **(1)**
(b) vertical transmission / from parent to daughter cells **(1)**
conjugation / transfer of plasmids between bacteria **(1)**

139. Inheritance

1 (a) carrier / heterozygous **(1)**
(b) Mother: Hb^S Hb^N
Father: Hb^S Hb^N **(1)**
(c) Hb^S Hb^S Hb^S Hb^N Hb^N Hb^N **(2)**
(**1 mark** for two correct)
(d) 0.25 **(1)**

2 (a)

	RF	Rf	rF	rf	
RF	RRFF	RRFf	RrFF	RrFf	(1)
Rf	RRFf	RRff	RrFf	Rrff	(1)
rF	RrFF	RrFf	rrFF	rrFf	(1)
rf	RrFf	Rrff	rrFf	rrff	(1)

(b) red + smooth, red + frilled **(1)**
white + smooth, white + frilled **(1)**
(c) 9 red + smooth : 3 red + frilled : 3 white + smooth : 1 white + frilled **(1)**

140. Linkage and epistasis

1 (a) sex linkage **(1)** (c) 0.5 **(1)**
(b) X^BX^b X^bX^b **(1)**
X^bY X^BY **(1)**

2 (a) BBEE BbEE **(1)**
BBEe BbEe **(1)**
(b)

	bE	bE	be	be	
be	bbEe	bbEe	bbee	bbee	(1)
be	bbEe	bbEe	bbee	bbee	(1)
be	bbEe	bbEe	bbee	bbee	(1)
be	bbEe	bbEe	bbee	bbee	(1)

(c) 0.5 **(1)**

141. Using the chi-squared test

1 There is no difference in the height of the plants in the shaded and unshaded areas. **(1)**

2

	O	E	(O − E)	$(O-E)^2$	$\frac{(O-E)^2}{E}$	
Shaded area	68	82	**−14**	196	**2.4**	(1)
Unshaded area	96	82	14	**196**	**2.4**	(1)
				Σ =	**4.8**	(1)

3 The χ^2 value of 4.8 is greater than the critical value at $p = 0.05$ **(1)** at 1 degree of freedom. **(1)** This means that the numbers of plants in the two areas are significantly different. **(1)**

4 availability of sunlight **(1)**

142. The evolution of a species

1 (a) natural selection **(1)**
(b) two populations become geographically isolated **(1)**
no interbreeding between the two populations **(1)**
different selective pressures in each area **(1)**
individuals who have the best adaptations survive and have offspring that share these adaptations, which differ between the two areas **(1)**
(c) sympatric speciation **(1)**

2 (a) predation by the small birds **(1)**
(b) Snails that are black or white are not camouflaged and are more easily predated by small birds. **(1)**
Grey snails are camouflaged and are less likely to be predated. **(1)**
(c) stabilising selection **(1)**

143. The Hardy–Weinberg principle

1 (a) $q^2 = 0.22$, therefore $q = 0.469$
$p = 0.531$, therefore $p^2 = 0.282$ **(1)**
percentage that are homozygous for the dominant allele = 28.2% **(1)**
(b) $p^2 + 2pq + q^2 = 1$, $2pq = 1 - (p^2 + q^2)$
$1 - (0.22 + 0.282) = 0.498$
percentage of the population that is heterozygous = 49.8% **(1)**
(c) 53.1% **(1)**

2 (a) q^2 is 1 in 10 000 = 0.0001, which gives $q = 0.01$ **(1)**
From the value of q, we can find $p = 0.99$, and so $p^2 = 0.9801$ **(1)**
$2pq = 1 - (p^2 + q^2)$, so $1 - (0.0001 + 0.9801) = 0.0198$
percentage of the population that are carriers for phenylketonuria = 1.98% **(1)**
(b) number of sufferers = population × q^2
= 24 000 000 × 0.0001 = 2400 **(1)**
(c) 1% **(1)**

144. Artificial selection

1 (a) large milk yield / volume **(1)**
long lactation period **(1)**
high milk quality **(1)**
(b) breed cows with the best characteristics with bulls whose female offspring had the best characteristics **(1)**
breed together the offspring with the desired characteristics, but not any offspring with undesired characteristics, and continue for many generations **(1)**
(c) in-vitro fertilisation **(1)** cloning **(1)**
artificial insemination **(1)**

2 (a) Preserved species are a source of genetic material and can be used to select for new varieties. **(1)**
(b) two from: will not be in pain **(1)**
will not have a shortened life expectancy **(1)**
will not suffer from difficulties with health **(1)**

145. Exam skills

1 (a) Haemophilia is inherited when only recessive alleles for haemophilia are inherited. **(1)**
The male sufferer has inherited an X chromosome from his mother and a Y chromosome from his father. **(1)**
The X chromosome carries the recessive allele for haemophilia but the Y chromosome does not carry any alleles for haemophilia **(1)**.
(b) mother is carrier, as son has haemophilia **(1)**
father doesn't have haemophilia, so probability is 50% **(1)**
(c) Haemophilia is a recessive trait; so two recessive alleles are required in order to inherit haemophilia. The recessive allele is carried on the X chromosome. Males have one X chromosome and females have two X chromosomes, so it is less likely that a female will inherit haemophilia compared to a male. The males in the diagram with haemophilia have inherited the recessive allele from their mothers, as neither parent has haemophilia. None of the females in the diagram have inherited haemophilia from their parents. In

order for a female to inherit haemophilia, her father must carry the recessive allele on his X chromosome, and would therefore have haemophilia himself. **(6)**

146. DNA sequencing

1 (a) Gene sequencing is the process of determining the order of the nucleotides in the DNA. **(1)**
(b) DNA polymerase **(1)**
adds nucleotides and dideoxynucleotides onto the template DNA in a complementary manner. **(1)**
(c) AACGGGCTCCCC **(1)**
(d) Dideoxynucleotides are labelled with a fluorescent dye. **(1)**
The computer uses a laser to read the fluorescent dyes at the end of each DNA fragment as it passes the end of the electrophoresis gel. **(1)**

2 (a) two from:
compare human gene sequences / genetic fingerprinting **(1)**
compare genomes of different species / examine evolutionary relationships **(1)**
research the genotype–phenotype relationships **(1)**
epidemiological research **(1)**
(b) DNA templates are attached to microbeads and many copies are made using the polymerase chain reaction. **(1)**
The templates can be sequenced in parallel. **(1)**

147. Polymerase chain reaction

1 (a) PCR is a method of making many copies of DNA. **(1)**
(b) Add DNA template to nucleotides, primer and Taq polymerase and add to thermocycler. **(1)**
Mixture is heated to 94–96 °C to separate the two DNA strands / denaturation. **(1)**
Mixture is cooled to 68 °C to allow primers to bind to the template / annealing. **(1)**
Mixture is heated to 72 °C to allow Taq polymerase to add nucleotides to the new strand of DNA / elongation. **(1)**
Process is repeated until required amount of DNA is obtained. **(1)**

2 (a) Traces of DNA from a crime scene can be amplified by PCR **(1)** in order to identify criminals **(1)** or samples can be used to show parent/child relationships. **(1)**
(b) two from:
too little DNA **(1)**
DNA damaged / fragmented **(1)**
DNA in sample contaminated **(1)**

148. Gel electrophoresis

1 (a) Restriction enzymes cut the DNA into smaller fragments. **(1)**
DNA fragments loaded onto the gel and electricity is passed through the gel. **(1)**
DNA fragments are negatively charged and so move towards the positive electrode. **(1)**
DNA fragments are separated according to size; large fragments at the top of the gel, smaller fragments at the bottom of the gel. **(1)**
(b) gene sequencing **(1)**
genetic fingerprinting **(1)**

2 (a) VNTR stands for variable number tandem repeat; they are sections of repeated DNA sequences. **(1)**
(b) Each person has a different number of VNTRs. **(1)**
Each person's VNTRs are in different positions in the genome. **(1)**
Restriction enzymes cut the VNTRs wherever they are found in the genome, so different numbers of sections of different length are formed. **(1)**
(c) People inherit half of their VNTRs from each parent. **(1)**
Half of the bands on a genetic fingerprint will the same as the mother and the other half will be the same as the father. **(1)**

149. Genetic engineering

1 (a) Genetic engineering is the use of technology to change the genetic material of an organism. **(1)**
(b) two from: making human insulin from bacteria **(1)**
making human blood clotting factors in goat's milk **(1)**
adding herbicide resistance genes to plants **(1)**
adding pesticide genes to plants **(1)**
(c) mRNA can be obtained from cells where the gene is being expressed. **(1)**
Reverse transcriptase can then catalyse the formation of a single strand of complementary DNA (cDNA) using the mRNA as a template. **(1)**
The gene can be synthesised using an automated polynucleotide synthesiser if the nucleotide sequence of the gene is known. **(1)**
The polymerase chain reaction (PCR) can use primers to amplify the gene from the genomic DNA if the nucleotide sequence of the gene is known. **(1)**

2 (a) A composite DNA molecule that contains genetic material from two different species. **(1)**
(b) Restriction enzymes / endonucleases cut specific sites either side of the gene of interest and a plasmid. **(1)**
The gene of interest can then be inserted into the plasmid using DNA ligase. **(1)**
(c) When the same restriction enzyme / endonuclease has been used to cut the gene of interest and the plasmid, the sticky ends of the gene of interest are complementary to the plasmid's sticky ends. **(1)**
The sticky ends overlap and can be joined by DNA ligase. **(1)**

150. The ethics of genetic manipulation

1 (a) two from: Bt toxin gene inserted into tobacco plants / maize **(1)**
genetically modified rice containing a gene to produce beta carotene **(1)**
genetically modified soya beans that are resistant to herbicide **(1)**
genetically modified plantains that contain more zinc **(1)**
genetically modified plants that are resistant to pests **(1)**
(b) When animals are genetically modified to produce pharmaceutical products in their milk or blood. **(1)**
(c) For (one from):
diabetics will receive human insulin **(1)**
genetically modified mice can be used to develop medical therapies **(1)**
people with emphysema will receive alpha antitrypsin **(1)**
Against (one from):
pain and discomfort caused to transgenic animals **(1)**
should not have patented animals as human property **(1)**

2 to prevent health problems associated with vitamin A deficiency **(1)**
genetically modified rice available in poor areas with a diet low in vitamin A and where rice is a staple food **(1)**

3 Farmers have to buy the seeds each year, instead of collecting them from their crop. **(1)**
Many farmers are poor and cannot afford to buy the seeds. **(1)**

4 Possibility that pollen from genetically modified plants could cross-pollinate with nearby plants. **(1)**
Possibility that other plants could incorporate the genetic modification into their own genomes, with unknown consequences. **(1)**

151. Gene therapy

1 (a) When the functional allele is added to the gametes **(1)**
all the somatic / body cells of the resulting offspring would contain the functional allele. **(1)**
(b) The functional allele would be passed on to offspring, with concerns about 'designer babies'. **(1)**

2 (a) severe combined immunodeficiency **(1)**
There is no functional adenosine deaminase / ADA allele, resulting in a non-functional immune system. **(1)**
(b) using liposomes as vectors **(1)**
delivering liposomes through an inhaler to be taken up by the lungs **(1)**

(c) It is difficult to get the functional allele into the right cells.
Need to make sure that the vector does not trigger an immune response. **(1)**
Need to provide the correct regulatory elements for turning the allele on and off at the correct time. **(1)**

152. Exam skills

1 (a) Heat the DNA sample to 94–96 °C to separate / denature the DNA strands. **(1)**
Cool the DNA sample to 68 °C to allow primers to bind / anneal. **(1)**
Heat the DNA sample to 72 °C to allow Taq polymerase to elongate / add nucleotides to the new strand of DNA. **(1)**
(b) gene sequencing **(1)**
(c) (i) Use restriction endonucleases / enzymes to cut the DNA of the mother, baby and possible fathers. **(1)**
Each lemur's DNA will cut into different length fragments. **(1)**
Run the DNA fragments on an electrophoresis gel to separate the fragments. **(1)**
Compare the positions of the bands, as the suspected father should have some of baby's bands in common. **(1)**
(ii) B **(1)**
because half of the baby's DNA fragments are the same as the mother's and the other half are the same as the father's **(1)**

153. Natural clones

1 two from: Strawberry plants reproduce using stolons. A new plant grows where a stolon touches the ground. **(1)**
Potatoes are tubers and produce sprouts on their surface that become new potato plants. **(1)**
Daffodils / bulbs grow an outer layer each year that grows into a new flower. **(1)**
2 asexual reproduction **(1)**
3 A cutting is taken from the parent plant. **(1)**
The cutting is dipped into plant 'hormones' / plant growth regulators. **(1)**
The cutting is placed into a new pot where it grows into a new, genetically identical, plant. **(1)**
4 All of the cloned plants would share the same desired characteristics, so a greater yield can be achieved.
Cloning plants that have been genetically altered to be resistant to pests means that fewer pesticides need to be used on the plants.
Many plants could be made from one parent plant, meaning that plants are produced more quickly than growing from seed.
Cloned plants are genetically identical and so are all equally susceptible to the same diseases.
Cloned plants will be less able to adapt to climate change as there is no opportunity for natural selection to act on variations, therefore there is less chance of survival of the species.
Using cloned plants leads to a reduced biodiversity in the habitat, which leads to fewer food sources for other species. **(6)**

154. Artificial clones

1 Tissue samples are taken from the parent plant. **(1)**
Tissue samples are placed onto a sterile agar plate containing nutrients and plant hormones / auxins. **(1)**
Samples grow into plantlets that are then planted into soil / compost. **(1)**
2 one from: no genetic variation **(1)**
all plants susceptible to the same diseases **(1)**
3 (a) making many identical embryos from one developing embryo **(1)**
(b) Each of the cells develops into an embryo. **(1)**
Each embryo is implanted into a uterus and they all develop into genetically identical animals. **(1)**
4 (a) one from: All animals have the desired characteristics. **(1)**
Large numbers of animals can be produced at the same time. **(1)**
(b) one from: no genetic variation **(1)**
health problems in cloned animals **(1)**

155. Microorganisms in biotechnology

1 the use of organisms in technology to make a useful product **(1)**
2 two from: brewing beer / wine making **(1)**
cheese / yoghurt production **(1)**
mycoprotein / Quorn™ production **(1)**
penicillin / insulin production **(1)**
3 Only a small number of temperatures and pH levels were used. The students should repeat this investigation using the same pH with more temperatures, and then the same temperatures with several different pH levels.
The opacity of the flask was estimated by holding the flask next to a standard and judging by eye, leading to inaccuracies. The opacity could be measured by using a colorimeter. Also, the opacity of the flask does not show the true number of bacteria in the flask. This could be measured by carrying out a serial dilution and then counting the number of bacterial colonies on a agar plate.
This investigation was only carried out once, so these results are not reliable. To be reliable, these results should be carried out at least three times to rule out any anomalous data. **(6)**

156. Aseptic techniques

1 (a) without microorganisms **(1)**
(b) two from: sterilising equipment **(1)**
hand washing **(1)**
wearing protective clothing **(1)**
2 (a) Batch fermentation:
nutrients and microorganisms are added to a fermenter **(1)**
and products are harvested at the end of fermentation **(1)**
Continuous fermentation:
nutrients are steadily added to the fermenter **(1)**
and products are constantly harvested **(1)**
(b) Unwanted microorganisms could compete with microorganisms in the fermenter **(1)**
and decrease the yield or produce a different product. **(1)**
Unwanted microorganisms could affect the health of workers. **(1)**
3 (a) batch fermentation **(1)**
(b) Penicillin is a secondary metabolite. **(1)**
Penicillin is only produced at a later stage. **(1)**

157. Growth curves of microorganisms

1 A lag phase **(1)** C stationary phase **(1)**
B log phase **(1)** D death phase
2 two from:
amount of nutrients / space **(1)** temperature / pH **(1)**
oxygen / water availability **(1)** toxic waste products **(1)**
3 (a) $7 \times 100\,000$ **(1)**
$= 700\,000$ bacteria **(1)**
(b) $700\,000 \times 80 \times 1000$
$= 56\,000\,000\,000$ / 5.6×10^{10} bacteria **(1)**

158. Immobilised enzymes

1 two from:
use of lactase to convert lactose to glucose and galactose **(1)**
use of glucose isomerase to convert glucose to fructose **(1)**
use of penicillin acylase to form semi-synthetic penicillins **(1)**
use of aminoacylase to produce L-amino acids for industry **(1)**
2 two from:
adsorption **(1)** – the enzymes are bound to a carrier made out of clay or carbon **(1)**
entrapment **(1)** – enzymes are trapped inside alginate beads or a gel **(1)**
cross-linking **(1)** – amino acids on adjacent enzymes are joined together through cross-links / glutaraldehyde **(1)**
3 can re-use enzymes **(1)**
the enzymes will not contaminate the product **(1)**
the enzymes will be able to tolerate temperature and pH changes **(1)**

4 Alginate beads containing glucose isomerase are placed into a column and glucose is poured through the column. **(1)**
The isomerisation of glucose is catalysed by the enzyme and the product / fructose is collected from the bottom of the column. **(1)**

159. Exam skills

1 (a) two from:
brewing beer / wine / baking bread **(1)**
cheese / yogurt production **(1)**
mycoprotein / soy sauce production **(1)**
(b) (i) 37 °C temperature **(1)**
pH 7 / neutral pH **(1)**
oxygen available **(1)**
(ii) no down time / fermenter is always in use **(1)**
high yield **(1)**
(iii) use aseptic techniques / keep fermenter sterile to prevent unwanted microorganisms growing in the fermenter and reducing the yield **(1)**
prevent harm to workers from any unwanted microorganisms **(1)**

160. Ecosystems

1 (a) the non-living components of an ecosystem **(1)**
(b) two from: light intensity **(1)**
water / oxygen availability **(1)**
pH of the soil **(1)**
air movement **(1)**
(c) competition between different species **(1)**
2 (a) one from:
intraspecific competition **(1)** predation **(1)**
interspecific competition **(1)** disease **(1)**
(b) As the birds eat the insects, the bird population would decrease **(1)**
due to reduced amounts of food for them to eat. **(1)**
(c) The tree would lose more water / produce less fruit / have fewer leaves, **(1)** due to increased transpiration in the hot weather. **(1)**
Bird and insect populations would decrease due to decrease in habitat. **(1)**
Bird and insect populations would decrease due to lack of food. **(1)**

161. Biomass transfer

1 (a) the rate at which plants convert light energy into chemical energy through photosynthesis **(1)**
(b) two from: energy lost due to respiration **(1)**
energy lost as egestion **(1)**
energy lost due to excretion **(1)**
as biomass that cannot be consumed (hair, bones) **(1)**
(c) a stage of the food chain **(1)**
(d) less biomass and less energy available at each successive trophic level **(1)**
2 (a) Remove the water from one cabbage by placing it into an oven. **(1)**
Burn one gram of the dry mass in a calorimeter to measure the amount of energy. **(1)**
Scale up for the mass of the organism and then multiply by the total population. **(1)**
Repeat the experiment two more times in order to eliminate any anomalies. **(1)**
(b) In order to calculate the energy content of an organism, the organism must die. **(1)** Each cabbage in the field will have a different mass, so it is difficult to extrapolate the data from one cabbage for the entire field. **(1)**

162. Nitrogen cycle

1 Atmospheric nitrogen is fixed into ammonium by nitrogen-fixing bacteria / *Rhizobium* / *Azotobacter*. **(1)**
Ammonium can be converted into nitrites by nitrifying bacteria / *Nitrosomonas* in the soil. **(1)**
Nitrites can be converted into nitrates by nitrifying bacteria / *Nitrobacter* in the soil. **(1)**
Ammonium / nitrates are taken up by plants and used to make amino acids / proteins. **(1)**
Plant proteins return to the soil through the action of decomposers when the plant dies and are converted into atmospheric nitrogen by denitrifying bacteria in the soil. **(1)**
2 (a) leguminous plants / bean family **(1)**
(b) Nitrogen-fixing bacteria in the nodules provide the plant with ammonium. **(1)**
The plant provides glucose to the nitrogen-fixing bacteria. **(1)**
3 Add fertilisers to the soil.
Add manure to the field. **(1)**
Grow leguminous plants in the field. **(1)**

163. Carbon cycle

1 (a) microorganisms respiring **(1)**
decomposers causing decomposition of dead matter **(1)**
(b) stored as parts of plants by photosynthesis
stored when land organisms die as fossil fuels / coal / oil / natural gas **(1)**
absorbed by the oceans **(1)**
stored in sedimentary rock when sea organisms die **(1)**
2 (a) greenhouse effect / global warming / increase in global temperature **(1)**
carbon dioxide in the atmosphere traps heat from the Sun as infrared rays **(1)**
(b) less available habitat due to sea levels rising / ice caps melting **(1)**
less available habitat due to flooding / desertification **(1)**
migration of animals due to habitat loss / loss of food sources **(1)**

164. Primary succession

1 Pioneer species break up the bare rock and add a thin soil to the surface when they die. **(1)**
Small plants and grasses grow on the thin soil and add humus to the soil when they die. **(1)**
The soil eventually becomes deep enough for intermediate species to grow. **(1)**
When the soil contains enough nutrients and is deep enough for climax species / trees to grow, this is called the climax community. **(1)**
2 two from: mountain avens **(1)**
willowherbs **(1)** mosses **(1)**
dwarf fireweed **(1)** dwarf willows **(1)**
other examples of pioneer species **(1)**
3 (a) deflected succession **(1)**
(b) curled leaves / tiny hairs in the underside of leaves with reduced access to stomata **(1)**
to decrease transpiration and conserve fresh water **(1)**
root nodules containing nitrogen-fixing bacteria / *Rhizobium* **(1)**
to overcome the lack of nutrients in the soil **(1)**

165. Sampling

1 (a) Sampling is a method of estimating the number of individuals in different species in a habitat. **(1)**
(b) The line transect is laid across the area to be sampled. **(1)**
The number of individuals of each species is recorded at intervals along the line transect. **(1)**
(c) two from: moisture content of the soil **(1)**
temperature **(1)** light intensity **(1)**
pH of the soil **(1)** mineral content of the soil **(1)**
2 (a) Marked and unmarked animals have an equal chance of being captured.
Marked animals have had enough time to intermix with the rest of the population before recapture starts. **(1)**
There is no emigration or immigration of animals. **(1)**
The population is stable / there are no births or deaths in the time between the first catch and second catch. **(1)**

(b) population estimate = $\frac{24 \times 27}{8}$ **(1)** = 81 **(1)**

166. Population sizes

The second finch species will initially have a very slow population growth/lag phase, while the finches learn about their new habitat. Then the population of the new species will increase rapidly/ log phase as the finches mate and produce offspring. During this stage, it is likely that the second finch species will outnumber the original finch species, and will outcompete them for food. The population size of the original finch species will decrease.
The second finch species will eventually maintain their population numbers/stationary phase, where the number in the population will fluctuate around a set point. This is the carrying capacity of the tree habitat. If the population rises above the carrying capacity, some finches will die due to lack of food, and the population will decrease back to the set point.
The number of tree spiders will decrease as the number of finch increase. As the tree spiders decrease in number, the number of finches will also decrease. This in turn, will lead to an increase in the number of tree spiders.

167. Conservation

1 (a) the active and dynamic management of a habitat in order to preserve biodiversity **(1)**
(b) economic: fishing / timber / tourism **(1)**
ecological: pollination / water cleaning / nutrient cycling **(1)**
(c) Resources are regularly replaced. **(1)**
Resources do not run out. **(1)**

2 (a) Selective felling removes some trees from an area.
Clear felling removes all of the trees on one area. **(1)**
Both methods replace the trees that are removed. **(1)**
Both methods decrease the biodiversity of the area, but clear felling decreases the biodiversity more. **(1)**
(b) Coppicing produces many thin branches that could be used as arrows. **(1)**
The same trees can be used to make arrows each year. **(1)**

168. Managing an ecosystem

1 Human activity / pollution / building / farming has damaged the environment. **(1)**
Endangered animals need protecting. **(1)**
Resources may run out / be unsustainable. **(1)**
Human life may be affected due to unusable land / no jobs / polluted water / polluted air. **(1)**

2 limiting human activity in the ecosystem **(1)**
protecting species within the ecosystem from hunting **(1)**

3 two from:
Galapagos Islands **(1)**
Masai Mara region in Kenya **(1)**
Antarctica **(1)**
Terai region of Nepal **(1)**
peat bogs **(1)**
Snowdonia National Park (UK) **(1)**
Lake District National Park (UK) **(1)**

4 Human activity / building / travelling in the ecosystem may cause pollution. **(1)**
People need to build houses / grow food. **(1)**
People need to have jobs / make money. **(1)**

169. Exam skills

1 (a) Graph should have a small increase (lag phase), followed by a steep increase (log phase). **(1)**
Graph should have a short plateau (stationary phase), followed by a decrease (death phase). **(1)**
(b) During the lag phase, the yeast are making proteins essential for growth so there is little increase in the population size. **(1)**
During the log phase, the yeast are rapidly dividing. **(1)**
During the stationary phase, the yeast are maintaining their population size / the same number of yeast are produced as die. **(1)**
During the death phase, more yeast die than are produced due to lack of nutrients / build-up of toxins. **(1)**
(c) (i) The population of *S. cerevisiae* would decrease **(1)** due to increased competition for food / space. **(1)**
(ii) interspecific competition **(1)**

AS paper 1

1	B **(1)**	8	C **(1)**	15	D **(1)**
2	D **(1)**	9	C **(1)**	16	B **(1)**
3	B **(1)**	10	C **(1)**	17	A **(1)**
4	B **(1)**	11	C **(1)**	18	C **(1)**
5	C **(1)**	12	A **(1)**	19	A **(1)**
6	A **(1)**	13	B **(1)**	20	B **(1)**
7	C **(1)**	14	A **(1)**		

21 (a) **(2)**

(b) straight chain with β 1-4 glycosidic links;
each cross linked to others with hydrogen bonds;
to form microfibrils;
giving strong multi-fibre structure (like rope);
β 1-4 glycosidic links not easily hydrolysed;
so fibres last a long time;
arrangement of microfibrils (gives an open structure to allow materials in and out of cell / ensures wall permeable) **(4)**
(NB: maximum of 3 marks for structural points)
(c) calculation of rates, as well as the correct maths
pH 3 = 0.3 ÷ 4 = 0.075
pH 4 = 3 ÷ 4 = 0.75
pH 5 = 1.2 ÷ 4 = 0.3
pH 7 = 0.5 ÷ 4 = 0.125
pH 10 = 0
y-axis scale appropriate and plots accurate

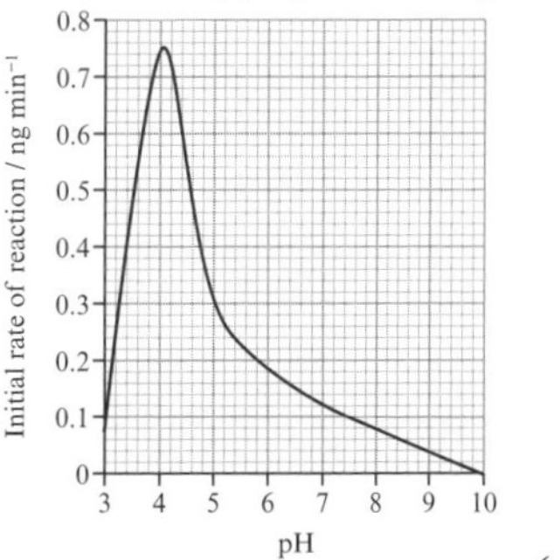

(4)

22 (a) AUCGGAUUUGCGUAC **(1)**
(b) nucleus **(1)**
(c) series of three nucleotides on the RNA complementary to anticodons on transfer RNA / tRNA
order of codons controls the amino acid sequence **(3)**
(d) Ile Gly Phe Ala Tyr **(2)**
(e) Vesicles from the rough endoplasmic reticulum carry the proteins along the cytoskeleton to the Golgi apparatus. The Golgi apparatus modifies the protein. Vesicles carry the modified protein to the plasma membrane. **(3)**

23 (a) (i) **A** spiracle **B** tracheole **(2)**
(ii) tracheal fluid removed (from tracheoles) into body fluid; increases surface area of tracheole (wall) exposed to the air; so more oxygen can be absorbed **(2)**
(iii) One mark for movement and one mark for explanation:
repetitive expansion and contraction of flight muscles; squeeze tracheal system; movements of the wings; alters volume of the thorax; as the thorax

volume decreases, air is pushed out of the tracheal system; some insects have developed this ventilation even further; by opening and closing valves in spiracles; air enters at front and leaves at the back of the tracheal system **(2)**

(b) (i) squamous epithelium **(1)**

(ii) walls one cell thick; covered in capillaries; capillaries one cell thick; short diffusion distance; many alveoli provides increased surface area for gas exchange; walls contain elastic fibres; recoil to force air out of lungs **(4)**

24 (a) diffusion; channel; exocytosis; active; ATP **(5)**

(b) (i) As the temperature increases, the percentage of pigment in the water increases, from 1% at 10 °C to 60% at 50 °C. From 50 °C to 60 °C the percentage of pigment in the water increases slightly. **(2)**

(ii) Pigment diffuses across the plasma membrane; through spaces between the phospholipids; as phospholipids move further apart due to increased temperature. **(3)**

(iii) The percentage of pigment in the water is estimated using colour standards, which is not an accurate method of measurement. **(1)**

25 (a) The absorbances of a set of samples of known concentrations of nitrates are measured by a colorimeter. The absorbances are used to make a calibration curve. The absorbances of the unknown samples are measured by the colorimeter. The calibration curve is used to calculate the concentration of the unknown samples. **(4)**

(b)

Sample	Absorbance	Concentration of nitrates/ mg ml^{-1}
A	0.1	4.5
B	0.4	18
C	0.2	9

(3)

(c) The concentration of nitrates is highest closest to the factory. Nitrates may have come from another source upstream of the factory / leaching of fertilisers from nearby fields / leaked sewage. **(2)**

AS paper 2

1 (a) peptide bond **(1)**

(b) (i)

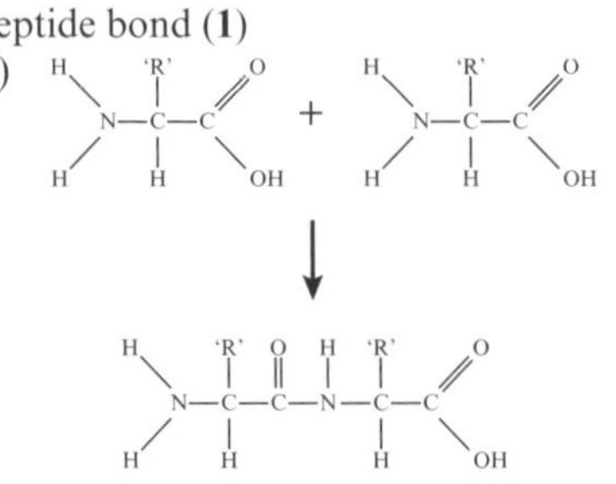

$$H_2N\text{–}C(H)(\text{'R'})\text{–}COOH + H_2N\text{–}C(H)(\text{'R'})\text{–}COOH$$

$$\downarrow$$

$$H_2N\text{–}C(H)(\text{'R'})\text{–}C(=O)\text{–}N(H)\text{–}C(H)(\text{'R'})\text{–}COOH$$

(3)

(ii) condensation reaction **(1)**

(c) Hydrogen bonds form between oxygen atoms and hydrogen atoms on nearby amino acids. Ionic bonds form between positively charged and negatively charged amino acids that are close together. Disulfide bridges form between cysteine amino acids that are close to each other. Hydrophobic and hydrophilic amino acids interact so that hydrophobic amino acids are located on the inside of the protein and hydrophilic amino acids are on the outside of the protein. **(4)**

2 (a) (i) −1% **(2)**

(ii) at 2.5% salt concentration because the same amount of water leaves the cells as enters the cells. The percentage change of the mass of the potato cells is 0. **(3)**

(b) (i) The potato slice decreases in mass by 3%. **(1)**

(ii) The plasma membrane comes away from the cell wall / plasmolysis. Cells become flaccid. **(2)**

(c) (i) the concentration of the sugar solution **(1)**

(ii) the original mass of the potato slices; surface area to volume ratio; temperature; volume of sugar solution **(4)**

(iii) Measure the mass of the potato slices before placing into sugar solutions.
Ensure that all potato slices are the same starting mass and surface area to volume ratio.
Use 6% sugar solution as well.
Replicate the investigation at least three times. **(4)**

3 (a) (i) The maximum volume of air that can be exhaled after a maximum inhalation. **(1)**

(ii) 4.5 dm^3 **(2)**

(b) (i) $(5 \div 20) \times 60 = 15$ breaths per minute **(2)**

(ii) $(1.5\,dm^3 \div 20) \times 60 = 4.5\,dm^3\,min^{-1}$ **(2)**

(c) (i) The student may have been unfit / sedentary.
The student may have suffered from asthma / COPD / emphysema / smaller body size. **(2)**

(ii) Make sure the spirometer mouthpiece has been sterilised.
Make sure that the carbon dioxide absorbent / soda lime still works. **(2)**

4 (a) (i) bacteria **(1)**

(ii) vaccination programme began; increased use of antibiotics **(2)**

(b) (i) Macrophages / neutrophils / phagocytes engulf bacteria.
A B lymphocyte that is complementary to the antigens on the bacteria is activated and produces antibodies. Antibodies bind to the antigen on the bacteria.
A T lymphocyte that is complementary to the antigens on the bacteria is activated, and produces killer T cells / helper T cells / regulatory T cells.
Helper T cells secrete cytokines to help B lymphocytes / stimulate phagocytes. **(5)**

(ii) active, natural immunity **(1)**

(c) No need to give BCG to all teenagers, as rates of TB infection are very low in the UK. **(1)** At-risk groups would still need to be given the BCG injection. **(1)**
If an infected person entered the UK, ring vaccination would be used to protect the population / all people in contact with the infected person would be vaccinated. **(1)**
The infected person would be treated with antibiotics. **(1)**

5 (a)

Domain	**Eukarya**
Kingdom	Animalia
Phylum	Chordata
Class	Mammalia
Order	Carnivora
Family	Canidae
Genus	**Canis**
Species	**lupus**

(6)

(b) eukaryotic cells; no cell wall / chloroplasts; multicellular **(3)**

(c) Both species feed on the same food sources.
If they both lived in the same territory it would lead to interspecific competition.
One species would suffer a decrease in population size. **(3)**

6 (a) (i) Herring has most similar proportions of all four nucleotides.
Mycobacterium tuberculosis and mangrove tree have the most similar proportions of nucleotides / have the highest proportion of C and G nucleotides.
Sea urchin has the highest proportion of A and T nucleotides. **(3)**

(ii) A binds to T and C binds to G. **(1)**

(b) (i) *Mycobacterium tuberculosis* (**1**)

(ii) It has the highest proportion of C and G nucleotides. C and G have three hydrogen bonds between them and A and T have two hydrogen bonds between them. It requires more heat to break three hydrogen bonds than is required to break two hydrogen bonds. (**3**)

A paper 1

1	B (**1**)	**6**	B (**1**)	**11**	B (**1**)
2	C (**1**)	**7**	B (**1**)	**12**	A (**1**)
3	C (**1**)	**8**	C (**1**)	**13**	B (**1**)
4	D (**1**)	**9**	C (**1**)	**14**	B (**1**)
5	C (**1**)	**10**	A (**1**)	**15**	D (**1**)

16 (a)

	Temperature/°C				
	10	20	30	40	50
net uptake of carbon dioxide in bright light (mg g^{-1} dry mass $hour^{-1}$)	2.4	3.9	2.2	0.3	0.0
release of carbon dioxide in the dark (mg g^{-1} dry mass $hour^{-1}$)	0.7	1.4	3.2	6.1	0.0
true rate of photosynthesis (mg g^{-1} dry mass $hour^{-1}$)	3.1	5.3	5.4	6.4	0.0

(**2**)

1 mark for correct values and 1 mark for correct decimal places

(b) respiration: 2.3; photosynthesis: 1.0, allow ecf; correct d.p. (**3**)

(c) respiration: 2.3 means rate doubles for each 10 °C rise in temperature; nothing is limiting the rate; rate is expected for chemical reactions in a test tube; do not accept higher rate than expected (**2**)
photosynthesis: 1.0 means rate does not change with the 10 °C increase in temperature; rate is limited (by CO_2 concentration); lower than expected for chemical reactions in a test tube (**2**)

(d) 0.00137 (a.u./arbitrary units)
Lose 1 mark for incorrect SF and lose 1 mark for inclusion of incorrect units (allow no units). 2 marks for correct answer with no working (**2**)

(e) CO_2 concentration, as when the CO_2 concentration is low the rate of photosynthesis is limited to a lower rate, regardless of the temperature (**1**)

(f) (experiment) C, as at a higher temperature and same CO_2 concentration the rate is increased. In B and A, CO_2 is the main limiting factor and in D there is no limiting factor in this experiment. (**1**)

(g) line should have a less steep gradient than experiment A. Rate levels off before or at the same light intensity as A (**1**)

(h) reduced CO_2 concentration: decreases rate of photosynthesis; source of carbon for production of (named) organic molecules; without CO_2 carbon cannot be fixed; Calvin cycle; CO_2 combined with RuBP (ribulose bisphosphate); to produce two GP (glycerate 3-phosphate)/carbon fixed; GP reduced to TP (triose phosphate); six turns of the Calvin cycle required to make (enough TP for) one molecule of glucose (**6**)

17 (a) ornithine cycle (**1**)

(b) A is urea; B is ornithine (**2**)

(c) deamination of / removal of ketone group from amino acids (**1**)

(d) ammonia is soluble; is highly toxic;
urea is less soluble; is less toxic (than ammonia); can be transported in the blood without affecting blood pH.
Max 1 mark if only mention of one of the substances with no comparison. (**2**)

(e) ethanol broken down to ethanal by ethanol dehydrogenase; ethanal broken down by ethanal dehydrogenase; into acetyl coenzyme A; which is used in the Krebs cycle; to produce ATP (chemical energy) (**4**)

18 (a) (i) A is spiracle; B is tracheole (**2**)

(ii) tracheal fluid removed (from tracheoles) into body fluid; increases surface area of tracheole (wall) exposed to the air; so more oxygen can be absorbed (**2**)

(iii) 1 mark for movement and 1 mark for explanation: repetitive expansion and contraction of flight muscles; squeeze tracheal system; movements of the wings; alters volume of the thorax; as the thorax volume decreases, air is pushed out of the tracheal system
Some insects have developed this ventilation even further; by opening and closing valves in spiracles; air enters at front and leaves at the back (of the tracheal system (**2**)

(b) (i) squamous epithelium (**1**)

(ii) walls one cell thick; covered in capillaries; capillaries one cell thick; short diffusion distance; many alveoli provides increased surface area for gas exchange; walls contain elastic fibres that recoil to force air out of lungs (**4**)

(iii) For both curves: percentage saturation with oxygen increases with increasing partial pressure of O_2;
at higher levels of O_2 saturation levels off. Ignore reference to individual curves, that is, less saturated than expected at low partial pressures and more saturated than expected at high partial pressures. (**3**)

(iv) 98% saturation of normal and 44% saturation of anaemic haemoglobin; half as much oxygen can be carried by anaemic haemoglobin (**2**)

(v) oxygen would be released from anaemic haemoglobin more readily at respiring tissues / greater oxygenation of respiring tissues (**1**)

(c) combination of CO_2 and H_2O produces carbonic acid / H_2CO_3; which dissociates into H^+ and HCO_3^-; H^+ combines with haemoglobin; to produce haemoglobinic acid; acting as a buffer; displacing O_2 from the haemoglobin; causing O_2 release to the surrounding tissues; high concentration of CO_2 in tissues results in more H^+ displacing O_2 and greater release of O_2 into the tissues (**6**)

19 (a) IAA and GA = 14 mm, control = 11.5 mm;
$((14 - 11.5) \div 11.5) \times 100 = 21.7\%$ increase.

(b) no treatment / no hormone (**1**)

(c) species of plant; age of plant; length of plant; concentration of hormone; AVP;
ignore length of stem section as is already controlled (**2**)

(d) (i) ethene / ethylene (**1**)

(ii) reduces labour costs; easier to transport while unripe; speeds up production (**1**)

20 (a) Type 1 is juvenile onset; type 1 is due to an inability of the pancreas to make enough insulin; thought to be the result of an immune response; type 2 is due to lack of / damage to receptors on the liver cells; so that they do not respond to insulin properly; can produce insulin but not enough (**3**)

(b) too much glucose in the blood; so not all is selectively reabsorbed in the proximal convoluted tubule (**2**)

(c) (i) made up (serial) dilution of known concentrations of glucose; added Benedict's reagent and heat; filtered samples; used colorimeter to measure absorbance (**3**)

(ii) line of best fit: curved line; anomalous result ignored or circled (**2**)

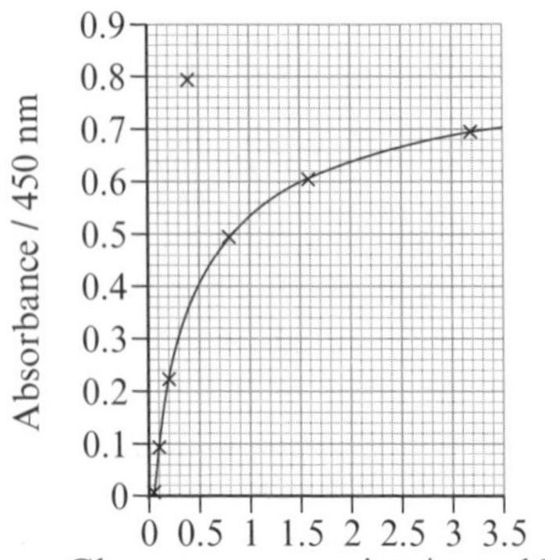

(iii) between 0.5 and 0.6 mmol L^{-1} of glucose; allow ecf from line of best fit **(1)**

(iv) within the normal range; no diabetes; no excess glucose in urine **(1)**

(v) repeat analysis of standards to eliminate anomaly; more intermediate concentrations of glucose; obtain more readings per standard and sample; repeat entire experiment (number of times); perform analysis of standards and unknown at same time **(2)**

(vi) use a glucose test strip; and one from: does not rely on dilutions for standard curve being correct; other substances in urine will not affect the reading; colour of the urine will not affect the reading **(2)**

A paper 2

1 C **(1)** 2 C **(1)** 3 D **(1)** 4 A **(1)** 5 C **(1)**
6 B **(1)** 7 B **(1)** 8 C **(1)** 9 D **(1)** 10 C **(1)**
11 D **(1)** 12 D **(1)** 13 B **(1)** 14 B **(1)** 15 C **(1)**

16 (a) no excess growth when unnecessary; will be one of the first flora to re-populate the area after a wildfire; plants without this ability may die out in the area as seeds may not be able to withstand the temperature / will have to wait for the temperature to fall again before germinating **(2)**

(b) secondary succession; of disturbed habitat; pioneer community; relatively few pioneering plants; such as *Caenothus* and others that germinate following high temperatures; some mosses, herbs and grasses; (after many years) fir and oak trees will become more prevalent; and decrease the number of shrubs and grasses in certain areas; eventually climax community will be re-instated
Ignore statements concerning primary succession. **(4)**

(c) Plants can absorb ammonia and nitrates; but not organic nitrogen; which must undergo decomposition; ammonification by bacteria. **(2)**

17 (a) mutation leading to more effective ADH4; in common ancestor
More effective ADH4: advantageous to break down alcohol; widens range of food that could be eaten; increased survival; better able to avoid predation; risk-taking behaviour could lead to finding more food / better shelter / migration; increased aggression to scare off rivals / win fights; more attractive as mates; more likely to survive; more likely to reproduce
Less effective ADH4: toxin build up could cause death; negative selection; more effective ADH selected for; over generations **(9)**

(b) 46% of East Asian descent have p^2 or $2pq$, so $p^2 + 2pq = 0.46$. The rest have $q^2 = 0.54$, therefore $q = 0.73$. $p + q = 1$ so $p = 0.27$ = ALDH2*2 allele
3 marks for correct answer **(3)**

(c) changes primary structure / sequence of amino acids; changes tertiary structure of protein; change in shape of active site; substrate no longer fits active site; no enzyme–substrate complex formed **(4)**

(d) 46% of; East Asians do not produce ALDH; so already do not convert acetaldehyde into acetic acid **(2)**

18 (a) autoimmune (disease) **(1)**

(b) sequence DNA of someone with SLE; and compare to someone without SLE; compare proteins produced **(2)**

(c) idea that T regulatory cells downregulate the immune system; negative selection; recognise cells as self; destroy lymphocytes with self-antibodies **(2)**

(d) binding to the antigen; complementary shape **(1)**

(e) (rheumatoid) arthritis; (type 1) diabetes; any correct answer **(1)**

19 (a) protoctist / protist **(1)**

(b) expulsive reflexes; blood clotting; inflammation; skin / blood clotting / wound repair / phagocytosis (only 1 mark as these do not protect against mosquito bite) **(3)**

(c) (i) macrophages; secrete cytokines; to signal to T lymphocytes; antigen-presenting cells; have direct interaction with T cell receptors **(3)**

(ii) plasma cells; produce antibodies; that are secreted; memory B cells provide immunological memory **(3)**

(d) memory cells produced in primary response; (so that) secondary immune response; is quicker; is stronger / more antibodies produced stronger **(2)**

20 (a) (i) correct $(n/N)^2$ values from table; 0.189 ; 0.81; full marks for correct answer with no working; allow e.c.f; minus 1 mark for incorrect rounding **(3)**

	individuals per m²		
Species	**Site 1**	**Site 2**	$\left(\frac{n}{N}\right)^2$
U	8	17	**0.057**
V	3	5	**0.005**
W	0	13	**0.034**
X	7	10	**0.020**
Y	2	17	**0.057**
Z	18	9	**0.016**
Total	**38**	**71**	**0.189**
D	0.64	**0.81**	

(ii) site 1; lowest biodiversity; less impact on the ecosystem **(2)**

(iii) two samples not representative of the whole site; could underestimate / overestimate the diversity; site 1 data not reliable; data could be inaccurate; should have large number of sample sites for each **(3)**

(b) (i) (gene) linkage **(1)**

(ii) parents both OoBb; gametes OB and ob; offspring OOBB, OoBb × 2, oobb
Award both marks if evidence of 3 : 1 ratio is clear. **(2)**

(iii) Agree: 3 : 1 ratio would be 150 orange/berberine to 50 yellow/no berberine; results not far off 3 : 1 ratio; only a few yellow/berberine and orange/no berberine; crossing over
Disagree: should only be parental phenotypes / equivalent; some yellow/berberine and orange/no berberine **(2)**

(iv) mutation; crossing over; AVP **(1)**

21 (a) reverse transcriptase **(1)**

(b) DNA polymerase; heat to break hydrogen bonds; cool to anneal primers; heat to make complementary strand; many cycles; DNA doubles with each cycle; AVP, e.g. further detail of temperatures / NTPs added **(5)**

(c) restriction enzymes (endonucleases); DNA ligase **(2)**

(d) insert another gene for particular feature, e.g. fur colour; extract DNA; use DNA probe to search for silk gene; AVP, e.g. details of DNA probe procedure **(2)**

(e) (i) embryos should be clones; same / genetically related surrogate(s); same mass / concentration of plasmid into each embryo; same delivery method; milk taken at same time; age of goats; same extraction technique; same technique for determining protein concentration; AVP **(3)**

(ii) natural variation; problem / mistake in data collection **(1)**

(f) (i) frameshift in triplet code / AW; early stop codon / shorter protein; different order of amino acids; AVP **(2)**

(ii) either post-transcriptional level; removal of introns / splicing; or post-translational level; phosphorylation / cAMP activation **(2)**

22 (a)

Order	Level	Name
7	Species	***gallus***
1	Kingdom	Animal(ia)
2	Phylum	Chordata
3	Class	Aves
6	Genus	***Gallus***
4	Order	Galliformes
5	Family	Phasianidea

Do not accept Gallus for species – must be lower case.
1 mark for correct order; 1 mark each for correct names **(4)**

(b) red junglefowl selected with desired characteristics; named characteristics; bred together; offspring with desired / named characteristics bred together; over many generations; kept separated from wild junglefowl **(4)**

A paper 3

1 (a) (i) In reference to secondary structure: hydrogen bonding; alpha helices / beta-pleated sheets.
In reference to tertiary structure: ionic bonds between charged, side chains / R groups / residues; disulfide bridges / bonds between sulfur (in cysteine residues), covalent bonding; hydrogen bonding between, side chains / R groups / residues; hydrophilic / hydrophobic / interactions between, side chains / R groups / residues.
Ignore references to primary structure. **(4)**

(ii) form the substrate binding pocket / bind to the substrate / phenylalanine; catalyse the formation of the product / tyrosine; aid in formation of the enzyme–substrate complex; correct ref. to the induced fit / lock-and-key mechanism **(2)**

(b) excess phenylalanine (in the bloodstream / liver); reduced tyrosine (in the bloodstream / liver); increased 5,6,7,8-tetrahydrobiopterin; reduced pterin-4a-carbinol-amine; reduced (metabolic) water production **(2)**

(c) (i) autosomal recessive **(1)**

(ii) 0.25 / 1 in 4 / 25% **(1)**

(d) enzymes / PAH trapped in a gel bead / matrix
Ignore any reference to adsorption, covalent bonding or membrane separation.
normal PAH used as a control; beads added to a column; indication of time allowed for reaction to occur
Independent variable: mutated / normal enzyme
Dependent variable: levels of phenylalanine / tyrosine in the product over a given time
Controlled variables: mass of enzyme / concentration of phenylalanine solution added / size of gel beads / size of cellulose fibres / timing of experiment / AVP
Accept concentration and volume of enzyme / phenylalanine.
Ignore concentration alone. **(5)**

2 (a) (i) 3519.37 (only one mark for > or < 2 d.p.) **(2)**

(ii) There is no difference between the levels of saxitoxin present in bivalve molluscs in summer and winter. **(1)**

(iii) There is a significant difference.
$p < 0.001$ / significance level of 99.9% / very significant **(2)**

(b)

Limitation	Improvement
no detail for random sampling technique given / whole population would not have been sampled	use metre taped to set out a grid and select random coordinates
no identification of species of bivalve mollusc / may be large variation between species	use a shore key to identify the species and analyse separately
only two seasons analysed	analyse other seasons (autumn and spring)

(2)

(c) Not valid: only one area sampled in each season, may not be representative; no indication of random sampling; species / genetic variation not taken into account
Valid: 60 organisms sampled (30 in each) so relatively large sample size; standardised protocol used to measure toxin levels; AVP **(3)**

(d) Level 3 (5 marks): An understanding that the toxin would block the channels AND a full explanation of how this would affect the action potential in the neurone AND the effectors.
Level 2 (3 or 4 marks): Some understanding that neuronal function would be decreased but insufficient detail of the role of the channels in propagation of the action potential.
Level 1 (1 or 2 marks): An attempt to describe the role of the channels in neuronal function. No appreciation of the effect on bodily functions.
0 marks: No response or no response worthy of credit.
Relevant points include: channels could be blocked by saxitoxin / saxitoxin could bind to channels preventing opening; channels normally allow Na^+ to diffuse into neurone; normally open at threshold potential to initiate action potential; local current would not activate action potential along neurone; no release of (named) neurotransmitters at presynaptic knob; no depolarisation of postsynaptic membrane; named effectors (muscles, release of hormones, e.g. ADH in osmoregulation, control of HR, control of breathing). **(5)**

3 (a)

Adaptation	Advantage
hairless	remain cool in hot environment
wrinkled skin	larger surface area for heat loss
	camouflage against desert
large protruding teeth	to dig tunnels
	prevents swallowing dirt when tunnelling
long whiskers	to sense vibrations in the tunnel / feel their way in the dark
short, thin legs	easy manoeuvring in narrow tunnels

both the adaptation and the advantage are required for each mark **(2)**

(b) action potential in sensory receptor reaches the presynaptic knob opening Ca^{2+} channels; Ca^{2+} causes fusing of vesicles containing SP with plasma membrane (PM) and exocytosis of SP; SP diffuses across synapse; receptors complementary to SP present in PM of postsynaptic knob; SP binds to the receptors on the postsynaptic knob **(2)**

(c) translation **(1)**

(d) (i) bar chart plotted correctly; reject histogram or line chart (bars must not touch); x-axis labelled appropriately (organism) and y-axis labelled 'relative hyaluronan (% of control)'; axis correctly scaled (linear); standard deviation error bars plotted above and below the mean **(4)**

(ii) Naked mole-rats have higher levels of HA than the other organisms. Mouse and guinea pig have similar levels of HA to one another.
overlap of error bars / standard deviations suggests no difference between mouse and guinea pig
Human significantly lower than all other organisms. **(2)**

(iii) humans have low cancer resistance and mice and guinea pigs have intermediate resistance; accept figures stated, e.g. humans are 90% less resistant to cancer **(1)**

4 (a) (i) A is tidal volume; B is vital capacity; C is total lung capacity; 2 marks for all three correct; lose 1 mark for each incorrect label **(2)**

(ii) Athlete X fitter than athlete Y; athlete Y may have exercised not long before analysis; athlete Y may have been more apprehensive; athlete Y may have found it difficult to 'breathe normally' with the apparatus; rates have a natural variation; athlete Y could be female / younger / smaller than athlete X **(2)**

(b) (i) oxygen taken into lungs, carbon dioxide removed from spirometer using soda lime **(1)**

(ii) A is 0.039 $dm^3\ s^{-1}$; B is 0.037 $dm^3\ s^{-1}$
1 mark for correct values; second mark only if 3 d.p. and correct SI units **(2)**

(iii) sex; body mass / height / BMI; heart rate; tidal volume **(1)**

(iv) suggests high rate of gas exchange; more O_2 to muscles for respiration; more CO_2 removed from tissues / blood; higher O_2 concentration in blood/ haemoglobin **(2)**

(c) pyruvate dehydrogenase complex is required for the link reaction; acetyl-CoA (acetate) is the starting point of the Krebs cycle; aerobic respiration will be reduced; less ATP produced; to activate muscle fibres; pyruvate build-up prevents further conversion of lactate to pyruvate; lactate (lactic acid) build-up; could have increased breathing rate compared to normal; reduced carbon dioxide production; reduced oxygen consumption; reduced diffusion of gases to/from (named) tissues/lungs; lower performance in one-mile run **(5)**

5 (a) bacterium / bacteria **(1)**

(b) sample in more than one area / all areas known to contain badger populations; set traps more than once in a particular area; mark the badgers so that the same badger is not sampled more than once; limit handling to avoid human scent on badgers; release badgers in same area as trapped; only take as much blood as is necessary to complete testing **(3)**

(c) (i)

Sampling area	Number of badgers with TB	Number of badgers without TB
near bTB farm	10	40
Expected	15% of 50 = 7.5	85% of 50 = 42.5
$(O - E)^2/E$	0.833	0.147
Chi-squared test value	0.980	

(3)

1 mark for correct expected values; 1 mark for correct test value (0.980); 1 mark for prevalence of bTB is not significantly different from what would be expected in the UK

(ii) non-bTB farms have a much larger sample size **(1)**

(d) primary defences; such as skin and mucous membranes; fever to inhibit bacterial reproduction; phagocytosis; by neutrophils/macrophages;
Specific immune response; cell-mediated immunity; T helper cells bind to pathogen and stimulate phagocytosis; humoral immunity; antibody production by B plasma cells; antigen–antibody complex **(5)**

Published by Pearson Education Limited, 80 Strand, London, WC2R 0RL.

www.pearsonschoolsandfecolleges.co.uk

Copies of official specifications for all OCR qualifications may be found on the OCR website: www.ocr.org.uk

Typeset by Kamae Design
Illustrations by Tech-Set Ltd, Gateshead
Produced by Out of House Publishing
Cover illustration by Miriam Sturdee

The rights of Kayan Parker, Colin Pearson and Rebekka Harding-Smith to be identified as authors of this work have been asserted by them in accordance with
the Copyright, Designs and Patents Act 1988.

First published 2016

19 18 17 16

10 9 8 7 6 5 4 3 2

British Library Cataloguing in Publication Data
A catalogue record for this book is available from the British Library

ISBN 978 1 447 98429 0

Printed in Slovakia by Neografia

Acknowledgements
The publisher would like to thank the following for their kind permission to reproduce their photographs:

Science Photo Library Ltd: Don W. Fawcett 4

All other images © Pearson Education